PENGUIN BOOKS

THE MYTH OF SOLID GROUND

David L. Ulin is the editor of *Another City: Writing from Los Angeles* and *Writing Los Angeles: A Literary Anthology*. He is a frequent contributor to the *Los Angeles Times* and *LA Weekly*; he has also written for *The Nation*, *The New York Times Book Review*, and *The Atlantic Monthly*. He lives in Los Angeles.

Praise for *The Myth of Solid Ground*

"*The Myth of Solid Ground* is enriched by Ulin's wide knowledge of writing on California, but it is more than just a gathering of ideas. This book is a subtle, personal, and adventurous exploration of what an immense natural phenomenon means in our culture at large and in the imaginations of men and boys traveling the state. . . . It is less the ultimate technical account of what an earthquake is, how it comes into being and what it does than a fascinating survey of what 'earthquake' means to some of the most intelligent and passionate people engaged in the studying of the subject."

—*Los Angeles Times*

"This is a book that gains as it grows. Readers will be impressed with both Ulin's prose and the artfulness of the book's construction. They will know more about seismology than they may have guessed, and they will certainly look differently at California and the landscape that shapes it." —*San Francisco Chronicle*

"[A] fascinating book . . . Ulin is drawn powerfully to almost every aspect of earthquakes and earthquake lore, perhaps because they elude our grasp almost entirely. . . . There are moments when times stops for all of us and eternity shows its teeth. That's the alternate universe Ulin is talking about, and he does a creepy, brave, elegant job." —Carolyn See, *The Washington Post*

"This is a wry, hugely original and, at times, scary exploration of the unstable boundaries between science and folk magic in California's culture of disaster." —Mike Davis

"Earthquakes are unique in their capacity to evoke primal fear, existential musings and spiritual epiphanies, according to Ulin. . . . Although readers will learn a considerable amount about seismology past and present, and about the myths and science of earthquake prediction, Ulin is far more interested in describing his quest for seismologically induced existential enlightenment."

—*Chicago Tribune*

"[A] dramatic work of journalism and reflection, of science and séance. . . . As quake-chaser Ulin follows the San Andreas, he essays on chance and likelihood, he puts his ear literally to the ground, he spends many hours in the seismologists' labs and he finds an answer in finding no answer: It turns out that what we don't know and why we don't know what we don't know about quakes still forms the heart of the story." —*The San Diego Union-Tribune*

"A fresh, well-written examination of the most primal of all destructive forces. . . . Ulin deftly captures that whistle-through-the-graveyard uneasiness, barely suppressed, whether he's taking us on a foreboding subway ride with his five-year-old son, Noah, or pausing to contemplate the San Andreas fault."

—*The Cleveland Plain Dealer*

"David Ulin puts his ear to the ground and hears every Californian's heartbeat. Wise, witty and sweetly humble, his book opens fresh lines of insight into our earthquake culture. Whether you live in this zone or not, this book is the best 'know fault insurance' you can own." —Marianne Wiggins

"Ulin offers an engaging survey of modern earthquake science and places it within the context of the loopy denial mechanism at work in most Californians' 'If we don't think about it we'll never have to face the Big One' attitude. This smart, engaging book will make you think twice about the ground beneath your feet."

—*Entertainment Weekly*

"An unusual book, part memoir, part science writing, and part examination of the boundaries between science and magical thinking. . . . For Ulin, the uncertainty of living in earthquake country is simply life's general uncertainty magnified many times."

—*Newsday*

"Ulin's brilliant prose recalls Charlie Kaufman dialogue, as he takes his audience on a wild drive across a beautiful yet doomed state, his mind buzzing with apprehension, with geological facts and with meditations on the themes of time, certainty, and faith."

—*Publishers Weekly*

"As a fourth-generation Californian, I've often wondered why Californians seem not to worry overmuch, if at all, regarding earthquakes. Having read *The Myth of Solid Ground*, I now understand the combination of ignorance, denial, and a willingness to play the odds that keeps Californians from psychological meltdown. A scary but necessary book!" —Dr. Kevin Starr

The Myth of Solid Ground

Earthquakes, Prediction, and the Fault Line Between Reason and Faith

DAVID L. ULIN

PENGUIN BOOKS

For Rae, who knows . . .

PENGUIN BOOKS

Published by the Penguin Group
Penguin Group (USA) Inc., 375 Hudson Street, New York, New York 10014, U.S.A.
Penguin Group (Canada), 90 Eglinton Avenue East, Suite 700, Toronto, Ontario, Canada M4P 2Y3 (a division of Pearson Penguin Canada Inc.)
Penguin Books Ltd, 80 Strand, London WC2R 0RL, England
Penguin Ireland, 25 St Stephen's Green, Dublin 2, Ireland (a division of Penguin Books Ltd)
Penguin Group (Australia), 250 Camberwell Road, Camberwell, Victoria 3124, Australia (a division of Pearson Australia Group Pty Ltd)
Penguin Books India Pvt Ltd, 11 Community Centre, Panchsheel Park, New Delhi – 110 017, India
Penguin Group (NZ), cnr Airborne and Rosedale Roads, Albany, Auckland 1310, New Zealand (a division of Pearson New Zealand Ltd)
Penguin Books (South Africa) (Pty) Ltd, 24 Sturdee Avenue, Rosebank, Johannesburg 2196, South Africa

Penguin Books Ltd, Registered Offices: 80 Strand, London WC2R 0RL, England

First published in the United States of America by Viking Penguin, a member of Penguin Group (USA) Inc. 2004
Published in Penguin Books 2005

10 9 8 7 6 5 4 3 2 1

Grateful acknowledgment is made to the U.S. Geological Survey (United States Department of the Interior) for the use of materials from its collection.

THE LIBRARY OF CONGRESS HAS CATALOGED
THE HADCOVER EDITION AS FOLLOWS:
Ulin, David L.
The myth of solid ground : earthquakes, prediction, and the fault line between reason and faith / David L. Ulin.
p. cm.
Includes index.
ISBN 0-670-03323-5 (hc.)
ISBN 0 14 30.3525 8 (pbk.)
1. Earthquake prediction—California. 2. Earthquakes—California—Psychological aspects. I. Title.
QE538.8.U45 2004
551.22'09794—dc22 2004041491

Printed in the United States of America
Designed by Carla Bolte • Set in Giovanni, with Gill Sans

But everything is like we think it is, don't you get it? Out of the million little things happening on this beach, you can only be aware of seven things at once, seven things at any given time. . . . We never really get the whole picture. Not even a microscopic part of it. . . . Our delusions are just as likely to be real as our most careful scientific observations.

—Denis Johnson

ACKNOWLEDGMENTS

When I first began *The Myth of Solid Ground,* I used to joke that I would either learn enough about earthquakes to stay comfortably in California or so much that I'd have to move away. As it happens, I came to the former conclusion, but it took a lot of help to get me there. Without predictors such as Jim Berkland, Charlotte King, and Zhonghao Shou—all of whom gave graciously of their time and imagination—I would never have thought to look beneath the surface of seismicity, while researchers like David Bowman, Antony Fraser-Smith, Tom Heaton, Susan Hough, Lucy Jones, Allan Lindh, Andy Michael, Evelyn Roeloffs, Charles Sammis, Christopher Scholz, Kerry Sieh, Paul Silver, and Ross Stein offered a context from which to navigate this ever-shifting world. I am grateful to all of them, as well as everyone at the Southern California Earthquake Center, the California Institute of Technology, and the United States Geological Survey. But mostly, I'd like to thank Linda Curtis, for giving me access to the X-Files and serving as a source of encouragement and information. Without her, this book would be a very different piece of writing, if, indeed, it would exist at all.

The Myth of Solid Ground started out as a cover story for *LA Weekly*, and I owe an enormous debt to my then-editor, Janet Duckworth, both for coming up with the book's title and for her willingness to take a chance on this idea. My agent, Bonnie Nadell, championed the project from the outset; as always, she remains a tireless advocate for my work. At Viking, Paul Slovak offered invaluable editorial advice, as well as calm and measured patience and support. I can't thank him enough for his insights, which invariably refined and improved the book.

More than anything, I'd like to thank my wife, Rae Dubow, and our children, Noah and Sophie, who put up with endless nights and weekends of working. This book could not have been written without their sustaining love.

CONTENTS

The Myth of Solid Ground

THREE APRIL EARTHQUAKES

Let me tell you about the earliest earthquake I remember. It happened in the spring of 1980, when I was eighteen years old and living in my first apartment, on Haight Street in San Francisco, with two friends from high school, a collection of Grateful Dead tapes, and a glorious sense of aimlessness, of being adrift in a magical universe, where virtually everything I confronted in my daily life could be construed to harbor a hidden message of some kind. Later that year, over the Fourth of July holiday, I would ask for a sign of God's existence; two days afterwards, a car in which I was a passenger went out of control on the 101 just south of Novato, slamming into a guardrail and rolling once, end over end, onto the highway shoulder, yet somehow leaving all five of us who'd been inside miraculously unhurt. I mention this neither to support nor debunk the God story, but simply as an illustration, to show the kind of boy I was, the things I thought about, the way I saw the world. It was a period in which I spent a lot of time considering connections, pondering synchronicity and the heady, if inaccessible, question of truth, awash in the quest for ultimate answers and the meaning that, I felt sure, was waiting, if only I could peel back the surface of the earth.

Among the more self-aggrandizing legends to swirl through San Francisco in the days I lived there was one claiming that the city was a modern re-creation of the lost kingdom of Atlantis; both places, or so the story went, were ringed by water and anchored by large white pyramids with red blinking eyes at their apexes, and both (here's the self-aggrandizing part) represented landscapes of enlightenment in a universe of human darkness, zones of fulfillment where people could exist as their most heightened, elemental selves. From the perspective of the present, I now see this story for the provincial fantasy it was, but I remain struck by just how often, during the spring of 1980, I happened to hear it, from people who didn't know one another, people who had nothing in common, whose definitions of enlightenment could never have encompassed one another's points of view. For one friend, it was a matter of mass reincarnation: San Francisco, she told me, was filling up with reborn Atlanteans—which, according to her scattershot cosmology, meant anyone who had ever been drawn to the city, or, in other words, nearly all of us. Another friend took things a step further, insisting that when all the Atlanteans finally reached San Francisco, the city would be destroyed by earthquake and tidal wave, just as Atlantis itself purportedly had been. The year this would happen, she told me, was 1982, although she couldn't explain why, exactly, other than to say she'd heard it somewhere, from someone else along the never-ending daisy chain of myth. When pressed, she'd shrug her shoulders and point at the fog-swept hills or the bulge of Mount Tamalpais, reclining like a sleeping Indian princess in the soft green distance past the Golden Gate Bridge, and say in a voice marked equally by wistfulness and wonder, "It makes sense when you think about it. This place is just too beautiful to exist."

In the midst of all this legend making, I was asked by another friend, a girl named Lauren, to spend Easter Sunday in Marin County, at her grandparents' house. Easter was not a holiday to which I paid much attention—too fetishistic, too stained by blood and obligation, not to mention that I was Jewish—but living three thousand miles from home, I missed the rootedness of family, the quality of belonging, of a past that extended further back in time than yesterday or last week. Because of that, when Easter finally rolled around, my roommates and I accompanied Lauren and her parents across the Golden Gate, through the rainbow tunnel at the north end of Sausalito, and up into the hills of Mill Valley, where her grandparents lived in one of those cantilevered California contemporary cottages, all blond wood and sheeted glass, with small, neat rooms like ships' cabins, and a redwood deck projected on four ridiculously skinny stilts a hundred feet above the slope of a shaded ravine. What impressed me about the house—what always impressed me about houses like that—was the sheer impossibility of it, the way it not so much occupied a physical space as hung suspended above one; airy, ethereal, less a residence than a promise or a prayer. Inside, however, the place felt far more tangible, recognizable even: the heavy floors and thickly braided carpets, the linen napkins and crystal goblets, the ease with which Lauren's parents and grandparents sipped drinks and talked of absent relatives, while we, the teenaged visitors, remained polite but slightly distanced, as if we didn't want to give ourselves away. More than anything, I found myself reminded of holidays at my own home, substantial gatherings marked by generations, by a continuity that made me feel connected to the world. It was as if, here in this improbable structure, I had stumbled across a piece of ground that turned out to be solid, rather than one that existed, almost literally, in thin air.

I don't remember very much about that Easter Sunday, just a few scattered impressions here and there. I know that we had lunch, and that, caught again in the pull of family gravity, I made a phone call to my parents; I also know that, during the course of the afternoon, Lauren and my roommates and I slid down the hill beneath the house and smoked a joint, shadowed by the weight of that deck jutting out above us like a canopy of dreams. It was still light when we got back to San Francisco, or so my memory dictates, even if, thinking about it now, that seems impossible to me. Either way, at the end of the day I was alone in our studio apartment, looking out the kitchen window at the unkempt slope of Buena Vista Park where it crawled up the other side of Haight Street, and talking on the phone. Somewhere in the middle of the conversation, I felt a slight pitch and yaw, like a hiccup in the floor beneath me, and the whole room started to rock, gently, even easily, as if the world had been cast on rollers and was being shaken by a giant hand. Our apartment was on the third floor of an old wood-framed structure covered with what looked like pale green linoleum, and as the beams in the ceiling began beating out a distinct rat-tat-tatting rhythm, I found myself facing a strange dislocation, as if I'd been caught between moments, between "normal" reality and a new territory I didn't understand. "What?" I said, and then I turned, and—this I recall as distinctly as if it were happening this very instant—noticed our collection of motley thrift-store coffee cups start to dance upon the shelves. In some strange way, it was like what would later happen in that rolling car, the idea that I was in the grip of something, that time had stopped and the simplest things—my name, my sense of self, the position of my body—had been suspended, cast aside. "Do you feel that?" I heard through the telephone receiver, but before I could answer, the shaking stopped, leaving in its aftermath something like total stillness: no birds chirping, no wind rustling, not the sound of breathing even, just the squeak, squeak, squeak of my chandelier as it slowed its swaying, and the fine, high hum of the phone connection buzzing in my ear.

Given who I was and how I saw the universe, you might think I'd have taken my first earthquake as portentous, as a harbinger of things to come. But the truth is that, as soon as the temblor was over, it didn't even take a second for reality to snap back into position, as if nothing extraordinary had occurred. Out the window, I could see cars moving up and down Haight Street, and people passing on the sidewalk; inside, I finished my phone conversation, and went on about my day. It wasn't until late in the evening, as I lay in bed waiting for sleep to find me, that I began to consider what all this might mean. By then, I'd heard—from someone, a roommate perhaps?—that the quake had been a 4.5, big enough to rattle nerves and windows, but too small, really, to do lasting harm. In that sense, it was more or less a momentary pinprick, a reminder of the ephemerality of existence, albeit easily dismissed. Yet there in the darkness, I couldn't help thinking about the way the quake had rumbled out of nowhere, then disappeared as quickly as it had come. It was as if, under the surface of this placid Sunday, there was nothing you could count on, as if, much like Lauren's grandparents' cottage, California itself existed in a state of elaborate balance, equally solid and insubstantial, between the quotidian landscape of daily living and the explosive possibility that lay beneath. Ever since I'd arrived in San Francisco, I had wondered, in an abstract fashion, what an earthquake might feel like, how it would move me, what might happen, how I would react. Now that I'd been through it, the experience seemed like an initiation rite. It wasn't that I felt settled, nor that I understood, in any real way, where I was. But I had been given something. If nothing else, I had an earthquake story to tell.

The earthquake story I've just told you didn't really happen. Or, at least, it didn't really happen like that. There was no Bay Area temblor on Easter Sunday 1980; in fact, the only noticeable quake anywhere in California was a 3.3 down at the southern end of the state, near the Salton Sea. According to the National Earthquake Information Center, much of the day's seismic activity—a cluster of 3.1s and 3.2s, with a 4.3 towards midnight to cap things off—took place a thousand miles or so north of San Francisco, in the vicinity of Mount St. Helens, which less than a month and a half later would explode with the force of an atomic bomb, showering the entire West Coast with a relentless rain of flat, gray ash. The quake I remember, most likely, was a 3.5 that struck across the bay, in Emeryville, the next afternoon, although it's also possible that I'm thinking about any one of the three small tremors that shook Hollister, in the Santa Clara Valley, throughout the following Sunday, or the 3.0 in Livermore a few days after that. One of the first things you learn when you move to California is that earthquakes are as common as breathing, as the beat of blood in your heart and lungs and temples, although most of them are so small, 1.8s and 2.2s, that they affect us, if at all, at a level below conscious reckoning, a shadow region where the boundaries between what we believe and what we know are rendered indistinct.

It's tempting to read my memory lapse as a product of that indistinction, of the tendency, two decades later, for specifics to slip against each other like the edges of a fault. Certainly, this would explain all those small discrepancies of detail, the differences in date and magnitude, my unsettled sense of time. To be honest, though, I'm not sure that's really accurate, since there's nothing fuzzy in my mind about either the earthquake or my visit to Mill Valley—or, for that matter, the uneasy way they coincide. Even now, I can feel the sudden onset of the shaking and see my apartment start its fluid swaying, just as I can recall staring up at the underside of Lauren's grandparents' deck and wondering how it could remain aloft. The more I think about it, the more these moments seem continuous, like two halves of an extended dream. In that sense, what matters, strictly speaking, isn't the so-called truth of my recollections; what matters is the way they add up to some larger narrative, a myth of wildness, of instability, which, in staking out a passage between disconnection and rootedness, tells me something about the way we live in California, even as it creates a context where my experience begins to resonate against itself.

How do we talk about earthquakes? How do we even approach them, let alone integrate them into our lives? More than half a century ago, the social critic Carey McWilliams laid out a model of the process, one that found identity in the turbulence of the land. "On the basis of their reaction to the word *earthquake*," he wrote, "Californians can be divided into three classes: first, the innocent late arrivals who have never felt an earthquake but who go about avowing to all and sundry that 'it must be fun'; next, those who have experienced a slight quake and should know better, but who none the less persist in propagating the fable that the San Francisco quake of 1906 was the only major upheaval the State has ever suffered; and, lastly, the victims of a real earthquake—for example, the residents of San Francisco, Santa Barbara, or, more recently, Long Beach. To these last, the word is full of terror. They are supersensitive to the slightest rattles and jars, and move uneasily whenever a heavy truck passes along the highway." For anyone who's spent much time in California, McWilliams's words ring with authority. What's more, once you've been through the cycle, you never lose that edge of awareness, that anticipation of the quake to come. For years after leaving San Francisco, I would tense up at the rumble of the New York City subway or feel a strong gust of wind shake the walls of a house in which I might be visiting, and experience what I came to recognize as a muscular memory, a clenching of both body and imagination. Little more than a month after I returned to California, in 1991, I was awakened one morning by the tribal drumming of beam against beam in the walls of my bedroom, a sound that told me, as viscerally as any shaking, that I was back in the earthquake zone.

This is a story that takes place within that seismic landscape. It's a story that begins and ends with an earthquake, a story about the way that, here in California, the soil we stand on can, without warning, turn as fluid as the sea. It's a story about how, in the face of all that motion, we evolve elaborate strategies of protection, strategies that help us get on with our lives. Some of these strategies are talismanic, like our unwillingness to forget the long, drawn-out seconds of the shaking, as if in survival there is an element of protection, a spell cast against the fault line rumbling again. Others are more practical, like the ritual of putting bottled water in our car trunks or stashing canned goods and emergency money by the door. As with most stories, there are two sides to this one, which, for want of a better frame of reference, we may think of as reason and faith. But for all their differences, both are after the same elusive something, which is to take the movements of the earth and endow them with a mythology by which they may finally, inevitably, make a kind of sense.

Earthquakes have always inspired such an air of wonder, a middle feeling between fear and disbelief. Almost twenty-five hundred years ago, Aristotle suggested that they were triggered by breezes trapped in underground caverns; when the winds blew, the earth shook, and the size of the earthquake was related to the force of the gale. Here in Southern California, the Gabrielino Indians told one another a different story—that the ground lay spread across the shells of enormous turtles, who would argue and then swim in opposite directions, making the earth shake apart. In Siberia, local legend was not dissimilar, if somewhat less picturesque: the world, or so it was believed, sat in the well of a giant sled, and every time the dogs who pulled it stopped to scratch their fleabites, a temblor would result. For ancient Peruvians, however, quakes were actually the footsteps of God himself, who periodically returned to earth to tally up the numbers of his children, even if, in the process, he killed many of those he had come to count.

From the perspective of the present, it's hard to frame these stories as anything other than folklore, which, of course, is what they are. Yet the more I consider them, the more intrigued I find myself, and not just from an aesthetic point of view. Earthquakes, after all, do strange things to our psyches, by shattering what may be our most widely held illusion, the inviolability of solid ground. Not only does this undermine our belief in the everyday stability of existence, it also offers, in a way I can't pin down exactly, evidence of an entirely different vision of reality:

odd juxtapositions, inexplicable happenings, situations that don't add up. At 4:31 a.m. on January 17, 1994—the moment of the Northridge earthquake—my wife and I both sat bolt upright in bed at the exact same instant, looked at each other, and said, in something close to unison, "Quake!" By the time the shaking kicked in, a split second later, we were already scrambling towards our bedroom doorway, where we huddled for the duration while outside, electric transformers blew, one after another, like giant flashbulbs, and a million car alarms sliced the eerie stillness of the night. Across town, a friend of mine, a long-lapsed Catholic, awoke at two that morning, and, seized with a dread she couldn't dissipate, began to pray, on her knees, fingers knitted, stopping only two and a half hours later, when the quake tore through her hillside home. On the one hand, incidents like these leave me skeptical; what are they if not coincidences, tiny blips of psychic radar, bits of something we imagine to reassure ourselves, something that isn't real? I'd be lying, however, if I said I didn't wonder about them, didn't stay awake nights trying to put together all the pieces, as if somewhere deep inside the moment an answer might be found.

All of this might seem like wishful thinking, but I prefer to look at it in terms of intuition instead. Even geologists acknowledge the role of suggestion, of serendipity, in their investigations; theirs is, by necessity, a conjectural discipline, in which verifiable information is hard to come by, and science blurs into what John McPhee calls "geopoetry," as in "where gaps exist among the facts of geology the space between is often filled with things 'geopoetical.' " Surely, geopoetry accounts for the current defining paradigm of plate tectonics—what else is it but a pure piece of poetry to see the earth as a loose collection of floating landmasses, ethereal as a dreamscape, a stew of rock and magma that connects and splits apart and reconnects with stately elegance, both solid and fluid at once? The same is true of deep time, geologic time, whose incomprehensible, even terrifying, distances are transformed by geopoetry, by the notion that, as one scientist says in McPhee's *Annals of the Former World*, "If you free yourself from the conventional reaction to a quantity like a million years, you free yourself a bit from the boundaries of human time." That the earth could exist in a state of constant evolution is as vivid a metaphor for the concept of a living planet as we're likely to come across, and it gives the entire arc of lithic history, and our finite, fleeting place within it, a grandeur as vast as that of heaven itself. With such a metaphor as a starting point, it may be the ultimate piece of geopoetry to imagine a system in which earthquakes can be read or reckoned with, in which there is a human logic to the geologic immensity, a way, in other words, to telescope time.

In his book *From Beirut to Jerusalem*, Thomas L. Friedman describes how, during the early 1980s, Beirut residents developed a set of complex psychological defenses against the random devastation they faced every day. "I rarely heard any Beiruti admit," he writes, "that the violence around him was totally capricious and that the only thing that kept him alive was callous fate—which was the truth. Instead, I would hear people say about a neighbor who got killed by an errant shell, 'Well, you know, he lived on the wrong side of the street. It is much more exposed over there than on our side.' Or they would say, 'Well, you know, he lived next to a PLO neighborhood,' or, 'He shouldn't have gone out driving fifteen minutes after the cease-fire started; he should have waited twenty minutes—everyone knows that.' " I've never been to Lebanon, but I understand instinctively what Friedman means. Here in Southern California, one friend of mine won't stop his car under a freeway overpass, no matter how irate other drivers get. Another spent the first six months after Northridge sleeping in her clothing, including socks and shoes. As for me, I don't have an earthquake kit—because somewhere along the line, I decided that, were I to prepare myself, it would be a way of tempting fate. By now, I can't remember how this particular superstition got started, but it doesn't really matter in the end. What's important is not the idea, but what it tells us, which is that to live on any kind of fault line, we need to believe in something. This is a story about myth, after all.

At 5:12 a.m. on the morning of April 18, 1906, San Francisco was struck by what is still considered the most destructive temblor ever to hit California, a quake that, along with the three-day firestorm it ignited, flattened more than five hundred blocks, more than twenty-five thousand buildings, in the very heart of the city, reducing what had been the West's most vibrant metropolitan center to a landscape worthy of Hieronymus Bosch. Looking at pictures of San Francisco from before and after, you can see the scale of the destruction, a destruction so complete, so overwhelming, that even now, nearly a century later, it literally takes your breath away. At the Cable Car Museum in San Francisco, one wall is covered by a large photograph of the intersection of California and Market streets taken barely a month before the earthquake, revealing a cosmopolitan cityscape, with tall buildings and streetcars, and masses of pedestrians decked out in suits and bowlers—in short, much the same vista you'd find in New York or London, Paris or Philadelphia. I've stood at that corner thousands of times, watching the cable cars climb the slope of California Street like brightly painted insects, immersed in the bustle of business, the warm light of morning, the feeling of a city on the move. Across the room, near the museum's entrance, an old wooden stereopticon showcases the other side of the earthquake, offering a photo flip-book of what one caption calls "Ruins as Far as the Eye Can See." In the most stunning shot, Nob Hill stands stripped and ravaged, elegant mansions gone as if they never existed, silent hillocks covered in rubble and ash. Such an image makes three-dimensional the scene described by an anonymous reporter in a combined edition of the *San Francisco Call-Chronicle-Examiner*, published the morning after the earthquake:

> Death and destruction have been the fate of San Francisco. Shaken by a temblor at 5:13 o'clock [*sic*] yesterday morning, the shock lasting 48 seconds, and scourged by flames that raged diametrically in all directions, the city is a mass of smouldering ruins. At six o'clock last evening the flames seemingly playing with increased vigor, threatened to destroy such sections as their fury had spared during the earlier portion of the day. Building their path in a triangular circuit from the start in the early morning, they jockeyed as the day waned, left the business section, which they had entirely devastated, and skipped in a dozen directions to the residence portions. As night fell they had made their way over into the North Beach section and springing anew to the south they reached out along the shipping section down the bay shore, over the hills and across toward Third and Townsend streets.
>
> Warehouses, wholesale houses and manufacturing concerns fell in their path. This completed the destruction of the entire district known as the "South of Market Street." How far they are reaching to the south across the channel cannot be told as this part of the city is shut off from San Francisco papers.
>
> After darkness, thousands of the homeless were making their way with their blankets and scant provisions to Golden Gate Park and the beach to find shelter. Those in the homes

> on the hills just north of the Hayes Valley wrecked section piled their belongings in the streets and express wagons and automobiles were hauling the things away to the sparsely settled sections. Everybody in San Francisco is prepared to leave the city, for the belief is firm that San Francisco will be totally destroyed.
>
> Downtown everything is ruin. Not a business house stands. Theaters are crumbled into heaps. Factories and commission houses lie smouldering on their former sites.

The two-tiered banner headline, in boldface capital letters, screams out:

EARTHQUAKE AND FIRE
SAN FRANCISCO IN RUINS

The San Francisco earthquake is a watershed in California history, an essential turning point in the narrative of the state. It represents, in many ways, the genesis of contemporary California, the origin of our identification of the place as earthquake country, as a fractured landscape of devastating possibility, where hope and terror may abruptly coincide. That's not to say there weren't significant California temblors before the destruction of San Francisco; Southern California was rocked twice in the nineteenth century by powerful earthquakes, while in 1865, and again in 1868, the Bay Area experienced tremors large enough that each was referred to, in its own time, as the "Great San Francisco Earthquake." During the 1800s, however, California was still too unsettled for earthquakes to have taken their place as part of the subterranean mythos of people's lives. What culture existed was pioneer culture, defined by a wholly different set of legends—those of the missions, of the gold rush, of the building of the railroads. When, on the morning of January 9,

1857, the San Andreas Fault slipped at Cholame, in Central California, triggering the 7.8 Fort Tejon quake, only two people along the nearly two-hundred-mile rupture died. At the time, California's population was not much more than half a million, its written history—its records, its newspapers—less than ninety years old. Earthquakes were considered, when they were considered, as localized, even isolated, events, with no particular relation to anything but themselves.

One afternoon, in her office at the Pasadena field office of the United States Geological Survey (USGS), seismologist Susan Hough showed me an 1890 history of California called *The Golden State*, which insists, with a kind of native pride, that "indeed, compared with the earthquakes of other times and countries, California's earthquakes are but gentle oscillations."

"Gentle oscillations?" I repeated, not quite sure I'd read it right.

"Yeah," Hough said, laughing, as she took the old leather-bound volume from me and slid it gently back onto her shelf. "I love that. It's like they're talking about an entirely different world."

Within this entirely different world, the 1906 San Francisco earthquake reverberated like . . . well, like an earthquake, an upheaval that turned into its own odd kind of fault line, separating the present from the past. It shifted perceptions, shattered preconceptions, suggested whole new topographies of iconography and fear. Partly, this had to do with the earthquake itself, as a discrete geologic event, which did so much damage, over such a large area, that it changed how people thought about the land. "Rupturing the northernmost 430 kilometers of the San Andreas fault from northwest of San Juan Bautista to the triple junction at Cape Mendocino," explains a USGS Web site on the subject, "the earthquake confounded contemporary geologists with its large, horizontal displacements and great rupture length. Indeed, the significance of the fault and recognition of its large cumulative offset would not be fully appreciated until the advent of plate tectonics more than half a century later." Stories abounded about the devastation. Throughout Northern California, ground surface ruptures of five yards were not uncommon, and at least one measured twenty feet. In Point Reyes Station, the 5:15 commuter train to San Francisco was shaken off the tracks just minutes before departure; a photograph taken at the scene shows the locomotive laid out on its side along the railbed, a crazy zigzag of cars splayed in serpentine behind it, like playthings swept aside by a giant hand. Just to the south, in Olema, it was reported that a cow had fallen into a ground fissure and been crushed to death, a tale repeated so often, and with such convincing vigor, that even Grove Karl Gilbert, one of the preeminent geologists of his era, believed it. From the perspective of a less hysterical moment, the crushed cow saga is clearly an exaggeration or, perhaps, as Philip L. Fradkin theorizes in his anecdotal history of California earthquakes, *Magnitude 8*, a practical joke played by a farmer with a cow to bury, who "made up a good story for the bothersome newspaper reporters and geologists." Either way, it's instructive, a three-dimensional example of the way that, when it comes to earthquakes, the boundary between truth and legend becomes vaporous and indistinct. If the ground itself can shift and rumble, buckling streets and buildings and throwing trains from their tracks, it's not much of a leap to think that the earth might swallow an animal or, for that matter, fall away entirely, casting California out to sea.

What's interesting is that, in the case of San Francisco, all these mythic subtexts coexist with a wealth of documentation, from films and photographs to a wide array of firsthand testimony, and at least two large-scale official reports. In that sense, the temblor stands as the first truly modern earthquake, in which the evidence, such as the images, is as important as the event. The more I think about it, the more appropriate this seems, since for all we know of the earthquake, there is plenty that eludes us still. I'm not talking now about the larger, geologic issues—how earthquakes work, what starts and stops them, how they spread and grow. No, I mean details far more basic, which exist on the most mundane, superficial strata of the event. The death toll, for instance, reported at the time, and for years afterwards, as around five hundred, was almost certainly higher, especially when you consider the destruction of neighborhoods like Chinatown, where the immigrant population was recorded only sporadically, if at all. (The latest casualty estimates are five to seven times that initial estimate, although even this may still be low.) Magnitude, too, remains open to conjecture; in 1958, Charles Richter—who, with Beno Gutenberg, developed the Richter scale at the California Institute of Technology in 1935—put the temblor's strength at 8.25 or 8.3, which became the commonly accepted figure, as well as the popular threshold for what we think of as the "Big One," the eight-plus-point catastrophe every Californian dreads. Two recent studies, however, at the USGS and Caltech, suggest a magnitude of 7.7 or 7.9, which only points out how little we can say for sure. If we can't with any accuracy determine how many people died or track the size of the earthquake, how can we determine anything? Even here, it seems, we're in the realm of geopoetry, where intuition, the ability to make connections, may be the most essential tool we have.

Nowhere is this more clear than in regard to the documentary evidence of the earthquake, which, far from offering a cohesive, comprehensive vision, gives us only glimpses, brief moments of focus that raise more questions than they resolve. On the surface, that seems a paradox; how, after all, can you argue with a film clip or interpret a newspaper photograph as anything other than itself? Yet if you read the eyewitness accounts and study the pictures, you find yourself experiencing a kind of inner earthquake, in which you're left suspended between past and present, between the chimera of solidity and the realization that everything around you could, at any minute, disappear. This is what I felt when I first moved to California, before my first earthquake, when I used to walk the streets of San Francisco and try to see through every sidewalk and building, as if, were I able to peer closely enough, some hidden truth might be revealed. If the artifacts of 1906 have anything to tell us, it's that beneath such façades lies chaos, and whatever meaning we uncover is our own. The pictures and stories, in other words, are not transparent—that is, they don't really illustrate what happened, in any way that helps us understand. Rather, they function more as portals of imagination, letting us think our way into the experience while simultaneously reminding us of its irretrievable distance, a distance fueled by time. This is only heightened by their photographic nature, which makes these images both real and alien at once. Looking at them, then, is less like examining a historical record than staring through a window into a parallel universe, recognizable but different, in which everything we know is upside down.

That, of course, is exactly how it feels to go through an earthquake, and this uncertainty, this loss of equilibrium, is what gives these pictures their power. In one photo, of Van Ness Avenue, a row of Victorian buildings have partly slid from their foundations, leaning against each other like sloppy drunks. In another, taken in the Western Addition, two women wearing bustle skirts and bonnets promenade formally, in the style of the period, along a street sheared to rubble, where the only standing structures are a few unsupported chimneys and walls. Perhaps the most famous picture is one by Arnold Genthe, snapped the day of the quake from the top of Sacramento Street, looking downhill towards the bay, where clouds of smoke billow up to the horizon, and groups of people stand or sit almost to the vanishing point, watching as the fire comes their way. It's a shot I've seen quite often, and what intrigues me are its odd, everyday touches, its glimmers of life exposed. Although much of Sacramento Street appears intact, one building's front wall has collapsed, and on the second floor, you can see a corner of what looks like someone's bedroom, where a bureau stands, undisturbed by the temblor, with a picture framed above it, still hanging on the wall. If there's a better metaphor for the randomness of devastation, I'd be hard-pressed to imagine one, and it seems all the more miraculous compared with the destruction of the city as a whole. Far more horrifying is an aerial photo shot from a tethered balloon five weeks after the earthquake, which yields a panorama of San Francisco—or, more accurately, of all that's left. Running like a scar down the center of the picture is a thin line of buildings rising up along Market Street to the dome of City Hall. In the foreground stands the Fairmont Hotel, one of the few identifiable landmarks, brand-new in 1906 and totally unscathed. On either side of that line, though, there's absolutely nothing, just an endless sprawl of flattened, wasted blocks receding to the picture's edge. You can see the rumpled rise and fall of hills and the orderly grid of streets; the roads are open, wreckage cleaned up and removed. Still, except for a few trees or pillars sticking up like silent tombstones, nothing breaks the view. It's not even a ghost town, it's just obliterated. Outside of downtown, San Francisco is gone.

In the face of such a photograph, you can't really tell whether what you're seeing was caused by the earthquake or the fire. But that just adds another mythic layer to the experience, another question that's left unresolved. On the one hand, it's a chicken-and-egg issue; without the earthquake, there would have been no conflagration. At the same time, the source of the destruction couldn't be more essential, in terms of thinking about what the San Francisco earthquake means. For all the accounts and images, the reports in newspapers across the country, one of the earliest myths to emerge in the weeks and months after the earthquake was generated by the civic leaders of San Francisco, who thought it best for the city's future to blame the disaster on the fire. It's not hard to see their thinking, since fire could happen anywhere: both Boston and Chicago had burned as recently as the early 1870s, well within the range of public memory, and if you look at pictures of the Boston fire, what you see is strikingly familiar: a ravaged downtown, flattened, scarred with rubble, in which a few solitary buildings stand like shocked survivors, presiding over a wasteland of devastation and fear. Earthquakes, however, were an entirely different matter—unpredictable, unpreventable, and (worst of all) seen as indigenous to California, a danger that could be avoided by living, working, *investing* somewhere else. Confronted with this, and its potential impact on their ability to rebuild the city, businessmen and local officials put their own spin on the situation, downplaying all mention of the temblor in favor of a new myth, in which the most important narrative was that of San Francisco rising from the ashes to be reborn.

Over the years, the shift in focus from earthquake to fire has come to stand for many things—the cynicism of the city's leadership, the naïveté of its people, the primacy of the booster spirit in determining our image of ourselves. Without question, it's a classic California story, in which reality, seismic or otherwise, becomes secondary to a kind of wish fulfillment, the idea that, if we just long for something hard enough, we can literally will it into being. Still, for all that makes me uncomfortable about such a notion, I can't help but see San Francisco's reinvention of its own history in terms more complicated, more psychological, than mere politics or economics, as the product of what we might call native optimism or, at the very least, healthy denial. There's a certain suspension of disbelief, after all, that comes with living in California, a faith that the inevitable will never happen, or that, if it does, we'll be able to keep it psychically contained. To manage this, we need the reassurance of a larger allegory, in which the details add up somehow, and our lives along the fault line are redeemed. You can write that off as a quintessentially Californian delusion, but I prefer to think of it as self-preservation, an ongoing process of finding order in disorder, of taking the random pandemonium of an earthquake and reconfiguring it to make unexpected sense. "With the earthquake and fire," David Wyatt writes in *Five Fires: Race, Catastrophe, and the Shaping of California*, "San Francisco began the immediate translation of the text into the myth. . . . The particular story that San Francisco told itself about the earthquake and fire was of a city coolly eyeing its own destruction, a city acting 'casual,' as Kathryn Hulme describes a man blowing drifting char from his hands, 'casual when you knew he wasn't feeling so.' " This could be said for all of us in California, where, beneath the shadow of a looming earthquake, we go about existence all the same. And late at night, when we recall the precarious terms of that existence, we reassure ourselves with the legend of San Francisco, how it did not vanish like Atlantis, but was, instead, reclaimed.

Ninety-four years, almost to the day, after the San Francisco earthquake, I'm driving down the seam of the Southern California coast, traveling the 5 from Los Angeles to San Diego, watching the edge of North America melt gently into the surf.

Somewhere to my left, east in the desert, the San Andreas is dissipating at its southern terminus near the Salton Sea; to my right, the Pacific sparkles ocean-blue and hazy, like the literal end of the world. It's about eight thirty on a Friday morning in April, and the 5 is empty—or, if not empty, exactly, then lonely, desolate, in that peculiar Southern California way. What I mean is that, here, just south of San Clemente, surrounded by the brown rolling hills of northern San Diego County, you get that wistful sense of drift, of distance, the disassociation that its detractors like to say has been California's deepest, most defining fault line all along. This is the California where Richard Nixon once walked the beach dressed in suit pants and wingtips, pondering his paranoia; the California where, in March 1997, thirty-nine acolytes of Heaven's Gate swallowed phenobarbital and vodka and lay down in their black Nike sneakers to join the comet Hale-Bopp as it swept past Earth. It is, in other words, an apocalyptic landscape, one where people live in disconnection, as unrooted as dust upon the surface of the world. Even the subtle sweep of the shoreline is deceptive; although it looks unspoiled and inviting, much of it belongs to Camp Pendleton Marine Corps Base, and has been fenced off and marked with warning signs. As is so often the case in California, nothing on this drive is what it seems.

The same could be said about many of us traveling this freeway; certainly, it's true of me. I, after all, am on the road this morning in pursuit of my own kind of illusion, an illusion with the power to shake the world. What I'm after is an earthquake, and not just any earthquake, but one human-made, manufactured, an earthquake generated in a lab. There's something utterly inconsistent—disassociated even—about such an intention, for one of the trickiest aspects of studying earthquakes is that they can't be re-created, but only considered after the fact. "The earth is a huge laboratory, and our experiments are few and far between," Tom Henyey of the Southern California Earthquake Center once told me in his office at the University of Southern California. "Our experiments are the big earthquakes, and we have to wait for them."

Today, however, I'm not waiting. No, for the first time, I'm *anticipating*, looking at the dashboard clock and measuring out the minutes, translating them into the miles I still have to go. At ten a.m., on the campus of the University of California, San Diego, a team of structural engineers and earthquake specialists will jolt a two-story wood-frame house constructed on a shake table—which is a large freestanding platform connected to hydraulic pumps that allows researchers to simulate the movement of an earthquake. What it will look like I can't imagine, beyond a few blank thoughts of falling roof tiles or the thunderous roar of exploding earth. Of the temblors I've experienced, all but one have found me at home, in the sanctuary of my bedroom or my kitchen or my living room, which means the blips of motion I've detected have been internal, constrained by walls and floors and ceilings, measurable within that private universe. The only time I was out during an earthquake was in early 1995, when a low-four Northridge aftershock rattled the plate glass windows of a restaurant where I was sitting with my wife and infant son. There were some gasps, a shriek or two, and a feeling like a long suspended heartbeat. Then the shaking stopped, and everyone returned to eating, as if nothing had disturbed our meals.

There is, of course, a major difference between that earthquake and the one I'm on my way to San Diego to witness, and it has to do with everything we don't usually get to know. In the restaurant, as with virtually every other temblor I've been part of, I was captivated most by uncertainty, an open-ended quality like being caught in something I could not control. This is the nature of earthquakes, their flux, their great immediacy, the way they start out of nowhere and then grow, or don't, according to some internal mechanism of their own. Once the shaking begins, you have no idea how it will finish, if the walls around you will crack and tumble, or if, after one long, breathless moment of elasticity, reality will snap back into place. Today, however, speculation isn't a factor; the shake test will be self-contained. Not only do I know the duration (about twenty-five seconds) and the magnitude (6.7, the same as Northridge, which this experiment is meant to simulate), I know what kind of damage to anticipate. "Some things may go," says Jill Andrews, the Southern California Earthquake Center director for outreach, whom I called a few days earlier to get details on the test. "But the structure will stay up." When I ask how she's so certain, she laughs and tells me there's another test planned six weeks down the road, which doesn't leave much time for rebuilding. Later, I learn that, at this second test, the shaking will be of a much greater intensity—so much so that the house may well come down.

I pull off the freeway and wind my way along the hilly ridges of La Jolla until I find UCSD. By now the sun is high, and it's a classic Southern California morning, the kind of day people move here for. As I park the car and wander across a narrow campus street to the Department of Structural Engineering, I see students walking in fluid clusters, riding bikes, discussing weekend plans. Again, I'm struck by the multilayered life of California, how we reside in the shadow of enormous forces, yet go about our business anyway. It's the same impulse that fueled the rebuilding of San Francisco, the sense of hope, of wishful thinking, that opens into myth. Standing on the UCSD campus, I feel like we are all participating in a consensual hallucination, a hallucination of stability and faith.

It is this, of course, that I am hoping to find here, although

I'm not sure I'd admit that, even to myself. At the most basic level, I want this shake test to reveal some thread of certainty, something to believe. My desire is as simple as the wish to stand both inside and outside at the same time, to experience and observe an earthquake without any of the messy entanglements that always end up getting in the way. Here—or so my fantasy tells me—I will be ready; here, after all these years in the fault zone, I will finally know what's going on.

That illusion of control disappears the moment I weave through the television trucks clustered along the roadside and walk down a slight incline towards the structural engineering lab. Almost immediately, I am struck by how unlike this is to what I imagined, how anticlimactic, how unfulfilled. Tucked into a loading area off the sidewalk, a small amphitheater of folding chairs is set up for the loose aggregation of professors, building contractors, and government officials who are this morning's intended audience. A lectern stands at the front, unoccupied for the moment, while in the background, the lab building looms like a soundstage. A huge open space, cavernous, it has no front wall or doorway, and is bounded by a metal ceiling maybe a hundred feet high. The floor is concrete, covered with electric cables, on which a gaggle of camera operators jockey for points of view. Large metal winches hang from overhead, and massive I-beams protrude everywhere. It's all so different from my expectations that I don't at first notice the structure that sits, behind a makeshift barrier of yellow caution tape, in a corner of the building, along one interior wall.

I call it a structure because it's not a house, at least not in any way I understand. In the first place, it's small. Not quite model small, I realize as I walk over to examine it, but almost—a double-decker shack, with two tiny chambers on the ground floor and a single room upstairs. It's too small to live in, certainly, despite the minimal, almost random furniture: a desk in the front room with a mock computer console, next to a bud vase that contains a single rose. Even if the place were bigger, I'd be hard-pressed to think of it as homey, for it's oddly incomplete, more an idea than an actual building, framed in two-by-fours and plywood, like what you might see at a construction site. There's no staircase to the top floor, just a large square opening, like the passage to a loft. Downstairs, the walls are skeletal, unfinished, little more than hollow forms. Lest this feel too vestigial, someone has installed several idiosyncratic civilizing touches, like the pair of matched ceramic window boxes that adorn the second-story windows, or the chandelier hanging from the upstairs ceiling on a long linked chain. It's a hybrid, this building, neither fully real nor completely an illusion, which is a pretty good analogy for how I feel, as well.

By now, the impromptu amphitheater is relatively full, and the low wall along the sidewalk is thick with students, who have come to see the building shake. As security clears the area in front of the platform, I retreat to a loading deck behind the TV cameras, looking for a clear line of sight. Once I've found one, I open my backpack and withdraw the camcorder I placed there this morning, framing the small house in the viewfinder's eye. Before I left Los Angeles, I told my son, Noah, nearly six and fascinated by the notion that the earth is, in some fundamental way, unstable, that I would record the proceedings so he could watch the shake test for himself. Over the years, I've seen plenty of earthquake videos, home movies of swaying streetlights and power lines, security camera footage that captures display racks of food and magazines tumbling to the floor. Still, like those old photographs of San Francisco, such images have always appeared less than real to me, not evidence but something far less tangible, subterranean dreamscapes loosed upon the world. Given my growing feelings of ambivalence, I'm not expecting this to be any different, but the truth is that I made a promise, and besides, you never know. So I dutifully start the tape and begin shooting, as if, were I to capture just the right progression of details, all the rough edges of belief I can't quite reconcile might suddenly slip into place.

The engineers check the system one last time, and we all stand silent and respectful through a few brief speeches—the usual self-congratulatory backslapping, as if the ribbon on a bridge were being cut. This, we are told, is the "first full dynamic earthquake test ever performed on a full-scale wood-frame building in the United States." The motion matches that recorded in Canoga Park, in the San Fernando Valley, during the Northridge quake. As to why it's so important, much of California's residential housing stock is of this wood-framed variety, which means that millions of real lives are at stake. As one official points out, "There have been ten significant earthquakes in California since the late 1980s, and they've caused 123 deaths and 13,000 injuries."

We hear three quick blasts of an air horn, and two revolving siren lights, one at each end of the little house, begin to turn. Lines of red slash the room like knife points, and, as the horn recedes, the table starts to move. All at once, the house slips side to side with a gliding motion, vigorous yet graceful just the same. Through my viewfinder, it looks like a photograph, with none of the abruptness of an actual quake. The chandelier sways, gently at first, then more violently, and briefly, I'm reminded of that earthquake in my San Francisco apartment; one of the window boxes separates and plunges downward, shattering on the ground. The strangest thing about the experience is the quiet—not the stillness of an earthquake, but a more industrial silence, as if we're listening to the thrumming of a well-oiled machine.

Then, as quickly as it began, the shake test finishes, and, except for the chandelier, swinging in decreasing arcs behind the upstairs windows, the house subsides into stasis once more.

There's a smattering of applause, but it feels less like awe than obligation, the polite response to a theme park show. Once it dies down, the dignitaries and observers break ranks and mingle, and the house is opened up. For all the force of the jolt, the building has suffered virtually no damage—no obvious structural dislocation, a minimum of cosmetic disarray. The vase has tumbled off the desk and landed on the plywood subflooring, but not only is it unbroken, its flower remains unbudged. Staring at it lying on the floor like a discarded plaything, I feel like I should take this as a symbol that, even in the midst of all the shaking, we may still come through unscathed. I live in a house like this, have lived in a house like this for the entire time I've been in California; I rode out the Northridge earthquake on the second floor of a duplex built in 1928 deep in the flats of Los Angeles, and when I moved, it was to another wood-frame 1920s house, single-story, which I consider less a construction than a vessel, a boat that rides upon the land. Many nights, I've walked the narrow halls and rooms of that home and thought about how it survived the 1933 Long Beach earthquake, and the Sylmar quake of 1971, and, of course, Northridge, always Northridge, the crucible that continues to fire the souls of almost everyone I know. And yet standing here, surrounded by the exposed joists and beams of this experimental structure, I find myself curiously unmoved.

The reason for this doesn't hit me until later, once I'm back home in Los Angeles, sitting at the dinner table with Noah, showing him the video I've shot. Because we don't have a VHS adapter, we watch the tape on the camcorder's tiny playback screen, huddled together like two kids in the dark. When the building starts to shake, Noah's face breaks into a wide grin; "Cool," he singsongs, syllable extended, stretching like the smile across his face. "Let's see it again," he insists after the tape is over, so we replay it, and then another time. As this final viewing fades to static, Noah sits in his wooden chair, pondering something. Watching him go still and silent, I wonder if the footage has upset him, if some small seedling of worry has begun to blossom in his mind. Except for that one quake when he was a baby, Noah's never been through a temblor, never felt the ground shift, nor heard the beating of the beams above his head. Yet before I can ask him what's the matter, he turns and tells me, "It's too bad you couldn't stand inside."

It takes a moment for me to figure out what Noah's saying, but once I do, a lot of things come clear. He's right, of course, it *is* too bad I couldn't stand inside the shaking house; this is the crux of the whole situation, the core of my unease. What I witnessed in San Diego has almost nothing to do with why I drove there, with what I *wanted*, with what I came to see. I went looking for some objective experience, some answers even, a way of separating myself from the sensation of an earthquake and examining it from the outside. Yet in doing that, I've only ended up removing myself from the very elements that make earthquakes resonate: their wildness, their disruption, all that roar and tumult, the way, in the throes of a tremor, everything—past and future, memory and anticipation, faith and context—fades away for the duration, leaving us in a strangely timeless present tense. It is this the shake test lacked, this quality of fear, of wonder, this *awe* in the Old Testament meaning of the word. To be in the building, though . . . then you'd feel it, even if you understood it wasn't real. There's a fine line here, I know, and I can't say with any certainty how this might have mitigated my dislocation, except that, one way or another, I would have been moved. Still, the more I think about it, the more I feel like I haven't observed anything but some elaborate parlor trick, an illusion of the most fundamental kind.

After Noah and his sister, Sophie, go to bed, I watch the video once more, looking for something, some angle of reflection to the world. There's nothing there, however; the images I've captured are less real to me than those of the San Francisco earthquake, where at least I'm seeing real people, real destruction, no matter how far removed. It reminds me of a Web site I once visited called *Virtual Earthquake,* where, by triangulating a set of seismographs, you can track the epicenter and magnitude of an online quake. Although ostensibly an educational tool, hosted by California State University, Los Angeles, this site, too, can't transcend an edge of trickery, even down to the certificate it offers, endorsing successful users as "Virtual Seismologists." Why mess with virtual earthquakes when, if you just wait long enough, you'll experience a real one? Why focus on epicenter and magnitude when, in the end, they're little more than abstractions that seek to quantify what can't be quantified? At best, it leaves you feeling empty; at worst, the emotions it brings up are considerably more difficult to pin down. The first time I tried the program, I bungled the calculations, ending up with an epicenter a hundred miles off the mark. What better metaphor, I recall thinking, for the inscrutability of earthquakes, their inability to live up to expectations, their refusal to play along? The second time, I figured out from looking at the maps that the temblor I'd been given was Northridge, which only made me more aware of the futility of the game. On my computer screen, the earthquake broke down neatly, into lines and numbers radiating outward from a map as flat and featureless as a sheet of colored glass. There was no solace in those figures, no mystery, no transcendence, nothing close to what I'd felt that January morning once the shaking settled down. I drew some small satisfaction from identifying this particular "virtual earthquake" correctly, but I never made my way back to the site again.

On the one hand, that's the limitation of fact-based thinking, the way information can't satisfy us—at least not information alone. No, for that we need poetry, *geo*poetry, a strategy for making unseen connections, for turning loose accumulations of detail into psychic landscapes, that makes sense of how we live. It's one thing, after all, to watch a wood-frame house rumble on a shake table; it's another thing entirely to live in one. It's one thing to recognize that the San Andreas yields a major Southern California earthquake every 140 years or so, which means that every quiet day we have here is another day we're overdue. It's another, however, to suggest that this in any way fulfills us, that it can match the mix of exhilaration and dread we feel each time the earth kicks into motion, a fundamental mystical quality that speaks to the very essence of our selves. If you go to the San Andreas, out in Redlands, say, where the fault runs fifty yards away from brand-new streets and housing developments, part of what you see is nothing, just a dry creek bed and vast empty distances, the only evidence of seismicity the soft contours of the mountains' alluvial fall. But if you stand there for a while, you notice the quality of the silence, the way that, besides the buzzing of an occasional insect, the quiet is as deep as the earth itself. It's a majestic silence, a mythic silence, the silence of the San Andreas as it waits.

I pop open a beer and walk the corridors of my small house, checking the corners for what the builders call stress cracks. Like most every home I've visited in California, we have them, adorning walls and doorjambs like small fault lines, jagged, waiting, hinting at the devastation underneath. Once, shortly after we'd moved in here, I asked the landlord about these cracks, about the way the back of the house, including the room I write in, sloped downward towards the backyard and a little to one side. Was the house just settling, I wondered, or were there problems with the foundation? If an earthquake happened, would I be able to get my family out? These are the most important questions you can ask in California, perhaps the only important questions, and yet they're questions that cannot be answered, questions that you have to take on faith. My landlord, of course, understands this; once I'd finished speaking, he simply stood there for a moment, looking up at the finely latticed lines that scar my office ceiling, and then, quietly, with a conspiratorial grin playing out along the corners of his eyes and mouth, said, "You know what I call this? It's California. That's the way it is."

He's right, my landlord, just as my son was right about the shake test: That's the way it is. That's the way we live here. That's the specter we confront. We sit in restaurants, park in underground garages, work in colossal office towers cast on rollers designed to minimize the shaking, although, really, no one knows. We drive our cars, we put our kids to bed, we revel in the shimmering sunlight, all the while trying to ignore the strange stratified nature of the present, where our most mundane activities play out between the residue of past disasters and the promise of disaster coming back. Sometimes, these hidden layers assert themselves when we least expect it, at the movies, say, or a baseball game. In the 1936 film *San Francisco,* which culminates with the 1906 earthquake, the final shot is a double exposure, in which the image of the new, rebuilt city is slowly superimposed over the burnt-out husk of the old. More than half a century later, during the 1994 season, the California Angels played in a stadium with a section of the outfield upper deck closed off, jagged where it had collapsed in the Northridge quake. I went to three or four Angel games that year, and I don't remember anyone ever saying a word about it, even though you couldn't help but notice, especially at sunset, when the broken edges of the structure purpled like ancient ruins in the long shadows of dusk. Eventually, the stadium underwent a major renovation, and when it was finished, the upper deck was gone. History erased, history rewritten, history eclipsed as if it never happened, except that, just like that ancient vision of San Francisco, history is always right at hand.

Not long after I moved to Los Angeles, someone told me that three hours before a major earthquake, the birds would stop singing, and, as at the San Andreas, there would be a profound stillness in the air. To this day, I think of that as a seismic barometer, a private source of reassurance, you might say. I don't know whether it works or not, since every temblor I've experienced has arrived at a moment of inattention, but when it comes to earthquakes, I'm willing to suspend my disbelief. I look for signs. We all look for signs.

THE X-FILES

The Southern California field office of the United States Geological Survey is located on Wilson Avenue in Pasadena, along the edges of the California Institute of Technology campus—a yellow two-story colonial house with black shutters that would not have been out of place on the set of *It's a Wonderful Life*. It's an unimposing structure, a little worn around the edges, like a well-loved family home. On the front lawn, a subtle marker identifies the building, while inside there's a comfortable clutter, a feeling of work in process, as if to mirror the flux and flow of the earth itself. To the left of the entry hall, in what at one time must have been a dining room, a large conference table sits surrounded by loose field kits and computer parts, and cardboard file boxes full of documents. A huge, three-dimensional map of Southern California takes up one long wall entirely, its dimpled surface covered with pushpins, each representing the epicenter of a different quake. As you stand there, seismologists pass through the halls in shorts and T-shirts, looking like nothing so much as middle-aged college students. Nowhere do you see any of the austere sterility popular culture associates with "science": no lab-coated technicians, no hushed technical language, no experiments beyond comprehension dealing with problems no layperson can understand. Across the street, in Caltech's Seeley G. Mudd Building of Geophysics and Planetary Science (or South Mudd, as it's commonly known), the lobby houses an Earthquake Exhibit Center, with a drum recorder seismograph and a computer console, featuring another map of California, which tracks every temblor in the region over magnitude 0.1, going back several weeks. Here, however, the only public display of earthquake culture is the collection of cartoons and tabloid clippings posted on the door of Linda Curtis's office, many of them curled and yellowed with age. In the center, a *Weekly World News* front page bears a "photograph" of the Devil as he emerges from a long tear in a San Fernando Valley street in the aftermath of the Northridge quake.

Curtis is, in many ways, the USGS gatekeeper, the public affairs officer who serves as a frontline liaison with the community and the press. Her office sits directly across the hall from the conference room, and if you call the Survey, chances are it will be her low-key drawl you'll hear on the line. In her late forties, dark-haired and good-humored, Curtis has been at the USGS since 1979, and in that time, she's staked out her own odd territory as a collector of earthquake predictions, which come across the transom at sporadic but steady intervals, like small seismic jolts themselves. "I've been collecting almost since day one," she tells me on a warm July afternoon in her office, adding that it's useful for the USGS to keep records, if only to mollify the predictors, many of whom view the scientific establishment with frustration, paranoia even, at least as far as their theories are concerned. "Basically," she says, "we're just trying to protect our reputation. We don't want to throw these predictions in the wastebasket, and then a week later . . ." She chuckles softly, a rolling R sound, thick and throaty as a purr. "Say somebody predicted a seven in downtown L.A., and we ignored it. Can you imagine the reaction if it actually happened? So this is sort of a little bit of insurance. If you send us a prediction, we put it in the file."

The file to which Curtis is referring is actually a loose collection of manila envelopes, accordion folders, faxes, e-mails, diagrams—everything, in other words, from single-sentence scraps of paper to the most elaborate presentations, bound in plastic report covers and featuring four-color art. Known throughout the USGS as the "X-Files," it's a fairly small conglomeration, which is surprising when you consider the hold of earthquake prediction on the public imagination, the fascination with signs and symbols, the unremitting desire to *know*. Prediction is just one endpoint of our tendency to mythologize earthquakes; there's little difference, really, between the traditional belief that temblors come from archetypal animals or giants, and the notion that they can be foretold if we just put the portents in their proper place. Both ideas represent attempts to explain the inexplicable, to make sense of the nonsensical, and both go back as far as history itself. Earthquake precursors (unusual premonitions, changes in groundwater, strange lights and fogs, and the reactions of so-called earthquake sensitives) appear in the literature of ancient Greece and Egypt; here in California, similar themes have held sway since the earliest days of the state. During the 1870s, the self-proclaimed Los Angeles prophet William Money foresaw the devastation of San Francisco by fire and earthquake, a vision that, Philip Fradkin notes, "was bound to be realized sooner or later, given the two Bay Area earthquakes in the 1860s and the fact that most of San Francisco had burned to the ground six times between 1849 and 1851." In May 1906, at a ceremony in the Sacramento Valley, a Wintun shaman interpreted the recent San Francisco earthquake as the beginning of a "great leveling" by which the world would be transformed. Predating all that is this 1869 account, by a traveler named Harvey Rice, detailing the relationship of earthquakes to electrical reactions in the earth's crust and the atmosphere. "We know," Rice declares,

> that electricity is an invisible force, active or quiescent, and abounds everywhere, in a positive or negative state. When the equilibrium has been disturbed, whatever the cause, it is certain that it will be restored by a like cause. The action of the electrical forces may or may not be instantaneous. The earth is said to be a great electrical reservoir, and so, in all probability, is the atmosphere; the one positive, the other negative, generally, or at points; yet always accumulating force, quietly, or violently, in the vain endeavor to restore a perfect equilibrium. Hence, we have thunder and lightning overhead and earthquakes under foot. The forces are the same. The one is a skyquake, the other an earthquake. The one would seem to be a substitute for the other, as in California, where they never have thunder and lightning, but are amply compensated by frequent earthquakes. For eight or nine months in the year they are favored with a bright sun and a cloudless sky. When the rainy

season commences, it brings with it violent electrical changes, resulting not in thunder and lightning, but in earthquakes. In this way, it may be presumed, the equilibrium is restored.

What makes Rice's observation so interesting is that, for all its quaint nineteenth-century supposition, it touches on theories that remain prevalent to this day. One of the most basic seismic myths, after all, is that of earthquake weather (although for us, the term signifies not rainy season, but its opposite—the hot, dry stillness that makes your skin crackle and your eyes dry out inside your skull), and it's equally common to hear talk of a link between earthquakes and electromagnetic disturbances, an idea that, over the years, has led to a wide range of predictions, at least some of which are impossible to dismiss. In 1989, a San Jose electronics salesman named Jack Coles called the Bay Area's Loma Prieta earthquake with the help of a homemade radio rig that picked up what he thought were "long-wave, low-frequency radio waves produced by the grinding of tectonic plates preceding an earthquake"; he used the same system to predict a 6.9 off the coast of Eureka in August 1991, as well as Northridge, which he forecast to the Associated Press twenty-four hours in advance. More recently, Kathy Gori, a Los Angeles sensitive, has run off a string of better than twenty successful predictions—with just a handful of misses—by relying on headaches that come and go a few hours before a quake. The key, Gori believes, is that her brain contains higher-than-average levels of magnetite, the mineral that helps bats and other animals orient themselves to the electromagnetic field of the earth, which enables her to function as a tectonic receiver, as it were. Even some members of the scientific community admit the possibility of a connection, among them Antony Fraser-Smith, a Stanford University geophysicist, who, like Jack Coles, stumbled across strange low-frequency radio signals in the days before the Loma Prieta quake. "When people start studying some sort of science," Linda Curtis suggests, "like, say, X-rays or space travel, none of it seems real or possible, and then you go forward a hundred years, and there it is. People who are sensitive to earthquakes may be sensing the whole thing Tony Fraser-Smith talks about, the shift in electromagnetics. Maybe people can be sensitive here."

For Curtis, all this is, in many ways, an exercise in wonder, in the limitless potential of the physical world. Although she describes herself as sitting "somewhere in the middle"—"I'm not totally dismissive, like a lot of people at the USGS, and yet I'm not going to jump in with both feet and say, 'This is the way to go' "—she's remarkably sanguine, even empathetic, about the predictors whose claims she has catalogued for more than twenty years. "The thing is," she says, as she gathers up a clutch of folders and ushers me across the hall, where she fans the files out like playing cards atop the conference table, "the people who do this genuinely believe in what they're saying, and in my heart of hearts, I would love for it to be true. I try it sometimes, between my husband and myself, like I'll say, 'I think there's going to be an earthquake tonight,' or 'I think there'll be an earthquake tomorrow.' And I have friends who do that, too."

The admission seems to make Curtis sheepish; once she's finished speaking, she grins and ducks her head for a quick moment, before taking a surreptitious look around. Then, she lowers her voice, as if, here beneath the radar screen of science, she and I have entered a conspiracy of faith. "Back in 1985," she recalls, her tone not much louder than a murmur, "I was living in Duarte, and it was a Saturday, and this door-to-door salesman came around. I don't remember what he was selling, but he came in and sat on my couch, and we started talking, and he asked what I did. So I told him, and he said, 'When's the next earthquake?'—which is what everybody says—and I said, 'Soon.' Two minutes later, while he was sitting there, there was an earthquake, and he said, 'Damn, I'm impressed.' The timing was so eloquent it was scary."

As Curtis talks, I'm reminded that the first time we met, in her office during the summer of 1998, she reported having predicted another small earthquake: "One morning, I said to myself, 'Next Tuesday, there'll be a 3.5 in Riverside'—and there was. I was so ecstatic, but I knew it was just random luck." Curtis, no doubt, was right about that, the sense of random luck, just as she's right about the timing of her Duarte quake. Still, left unspoken is the idea that such luck, such timing, may touch on something larger, some need for connection—if not within the earth, then within *us*. Once, not long after our initial meeting, Curtis told me that her interest in prediction was at least partly aesthetic, that the X-Files were compelling precisely because the documents gathered there were "interesting, creative, artistic. Everyone has their own ideas, and they're all distinct." Issues like these speak to the very heart of prediction, which is, at the most basic level, a matter of supposition, of suspending your disbelief. It is, in other words, a matter of geopoetry, the inexplicable territory of the imagination where fact and fantasy, desire and intuition, coincide.

Of course, the issue with geopoetry, I realize as soon as I dig into Curtis's X-Files, is that it's so vague; it can encompass anything. And anything is what I discover in this collection of folders and documents, an array of evidence certainly, although evidence of what, exactly, I'm not sure. Yes, many of these predictions operate out of intuitive, even poetic impulses, but it's a strange kind of intuition, and immersing in it is like taking an excursion into some uncharted interior wilderness, a region where nothing is as it seems. Once Curtis sets me up at the far end of the conference table and heads back to her office, I page through the various materials as if they were the individual chapters of a speculative novel, a book defined by its own lack of definition, by the way these random notes and notions yield one to the next with all the logic of a dream. Through the open windows, I can hear birds chirping (no earthquake for at least three hours), while from out in the hall, voices drift in and out of range with a monotonic hum. Here, though, I have crossed over into another landscape, where people go by strange sobriquets and nicknames, and their missives are often less forecasts of geologic activity than of the seismic shifts and ripples of their own elusive minds. There's the "Master of Disaster," whose faxed admonitions are surprisingly cheerful: "According to my calculations," reads one dated May 11, 1995, "conditions are favorable for producing an earthquake in the vicinity of Anchorage, Alaska, some time on May 12 or May 13, if ground energy fields line up properly. This seismic update brought to you by: The Master of Disaster. Have a nice day." There's Edward G. Muzika, who, on September 4, 1995, predicted a "magnitude 7.0 or greater quake, with an epicenter near Rancho Cucamonga" before the end of 1995, and, when that earthquake didn't materialize, was never heard from again. There's the anonymous predictor who warned of an 8.6 along the New Madrid Fault near Blytheville, Arkansas, on June 4, 1993, at 7:59 a.m.—a forecast so ridiculously specific even other predictors deride it as the work of a charlatan or a fool. And then there's John J. Joyce, a seventy-eight-year-old retiree from Dallas, Pennsylvania, who has sent in literally dozens of letters, most of them written in a cramped, compulsive scribble, spelling out a theory that involves something he calls "earthquake strains"—although what these are, and how they work, remains lost in the tangled circuits of what he admits is "a warped mind (mine)."

As much as anything in the X-Files, Joyce's letters exist not alongside reality but perpendicular to it; that is, while they open sensibly enough, they soon take a ragged turn and veer completely off the map. "It must be vacation time. We are having difficulty getting our mail on time," begins a May 18, 1998, missive, before it moves on to the discrepancies between Joyce's "research" and the state of current scientific understanding, the difficulty of "re-educat[ing] our geophysicist [*sic*] whose education had been over concentrated on plate tectonics as their answer to everything." For Joyce, the insistence of contemporary geologists on looking at plates and fault lines obscures the real cause and pattern of earthquakes, which have to do with electrical currents that can be traced by drawing lines across the surface of a globe. How he knows that . . . well, this is where Joyce's tenuous grip on logic deserts him, and his correspondence becomes a study in the chaos of the human mind. "There was one geophysicist," he writes,

> who said an earthquake could never be detected because they would never be able to tell where plates would shift. What he was saying made sense to me. I tackled it from a different direction. I had a twelve inch globe and marked it with crayons where I thought an earthquake would occur. I marked ten areas where I felt our earthquakes would occur. I was half right. The ten quakes did occur, but they were all out of sequence. I had been using a direct current to locate them. From that I concluded that I had the trigger that caused quakes, but had to find what was causing them to be out of sequence. To do that I had to call on my long term memory. I had read one time that there was very little eroded materials on the ocean floor. I figured that the eroded materials would have to travel under the deep part of the ocean where it would be subjected to extreme pressure and thus turned back into rock. This rock

> would be subjected to run into a larger land mass where it would again be subjected eventually to more erosion and be on its way again which would be the reason that it eventually came to rest on the California Coast. Eventually it had settled in the Ross Sea area of the Antarctica where its eroded materials was then sent North through its Pacific Current and the California Coast. Therefore instead of a D.C. current there became an A.C. current. It was then that the sequence of the quakes fell into place.

Taken at face value, there's little you can do with such a statement, little that it tells you, little sense it makes. My initial reaction is to chuckle, but it's not long before humor is displaced by a certain edge of desperation, desperation at Joyce's need, at his confusion, desperation at my own voyeuristic distance from the circles his mind makes. It's sad, after all, unsettling, this image of an elderly man crayoning lines across a globe and then announcing that they somehow enable him to predict when and where the ground will shake. It's like looking at the end of something, like watching the collapse of rational thought. At the same time, I can't help wondering where this leaves us, what hints or clues it may provide. The more I read, in fact, the more I start to see Joyce as a signifier, interesting—important, even—not for his ideas so much as his ideation, the act of theorizing itself. It's not that he has tapped into anything, although I'm struck by his attention to electricity, by the way that, sitting alone in Pennsylvania, sharing no apparent contact with either seismologists or predictors, he's stumbled across a common thread of earthquake forecasts, albeit in a way that bears no real relation to anything else. What makes him compelling, though, is his desire to frame his work in terms of geological "research," a large, empirical order in which his speculations might add up. He is after a belief system, a cohesive (if not coherent) way of thinking, one that has, for him at least, a scientific basis, that taps into the truth. The deeper I get into the X-Files, the more I recognize this kind of logic, the notion that, for their creators anyway, the theories here are not random but rational, seeking to establish common ground between mythology and fact.

It's tempting to dismiss such posturing as contrivance; how better to suggest you should be taken seriously than to argue that you're talking about reality, not conjecture, especially when it comes to quakes? Still, the fascinating thing about the X-Files is how ingenuous, how innocent, so much of its odd documentation seems. That can be a tricky idea to navigate, especially since the vast majority of predictions have been proven wrong by hindsight. Nonetheless, it raises questions about the nature of belief and credibility, which come into stark relief each time I stumble across a document that *does* seem to have something to it, like one dated December 6, 1996, calling for a 6.0–7.0 along the west coast of Mexico within a month (a 6.8 hit just off the coast of Michoacán on January 11, 1997, only six days late), or the elaborate booklet, complete with maps and graphs and color illustrations, prepared in late August 1993 by an "aero engineering design[er]" named Kenny Rogers, forecasting a giant Southern California earthquake sometime in January or February 1994. Sitting at the conference table, trying to decipher Rogers's charts and calculations, I keep thinking about Northridge, about the joists and wood beams beating against one another in the darkness of that January night. Is what I'm looking at now a fluke or could it be a warning? Is it possible that Rogers might have known?

Such speculations push the vagaries of earthquake prediction to a whole new level, not least because, despite its apparent accuracy, Rogers's prediction remains illusory at best. The temblor he forecast was an eight-plus point along the San Andreas, with an epicenter at Cajon Canyon, while the Northridge quake measured 6.7, and took place on a previously unknown thrust fault in the San Fernando Valley, sixty miles to the west. Still, if nothing else, the coincidence of timing brings about an almost physical shock that makes me wonder what I'm seeing, as if I'm digging at the roots of something not fully developed, in which discrepancies of magnitude and location may someday be resolved. On the cover of Rogers's booklet, there's a primitive photo illustration that portrays a highway rupture in which trucks go sliding down embankments, and the multiple lanes of Interstate 15 split and separate like seams. I can't see that picture without thinking about a series of news photos taken in the first hours after Northridge at a freeway interchange south of Newhall, where the Antelope Valley Freeway collapsed onto I-5. Much like Rogers's illustration, these pictures show a highway torn to pieces, one long slab of overpass broken off and down across the lower roadway, cars crashed or discarded at crazy angles, and the body of Clarence Wayne Dean, a Los Angeles motorcycle cop killed when the freeway disappeared beneath him, under a white sheet near his bike. That's a gruesome set of images, but what's striking is the odd way that it mirrors Rogers's own imagination, even down to the big rig left jackknifed and abandoned on another overpass.

Here, however, the split between science and supposition opens like a deep ground fissure, a vast and insurmountable seismic divide. The USGS has little interest in conjecture; even Linda Curtis says that, while some people may be onto *something,* it's nebulous, not really useful, while the geologists will tell you that Rogers is wrong, pure and simple, that the quake he forecast didn't happen, that there is no relationship, no confluence, that coincidences are a waste of time. Perhaps no one is as vehement on the subject as Lucy Jones, the Pasadena office's scientist-in-charge and longtime spokesperson, who is best known for the early morning press conference after the 1992 Landers earthquake at which she spoke to reporters while holding her own tired and frightened toddler in her arms. More than anything, Jones believes, this image has made her a lightning rod for various earthquake obsessives, as if she were a benign, maternal figure who might listen to their seismic fantasies—or, as she puts it, with something between a smile and a grimace, an "earthquake mom."

On some strange level, Jones is right about that, perhaps more deeply, even, than she's aware. Most of the letters sent to the X-Files *are* addressed to her, and many assume a revealing tone of familiarity, calling her "Lucy" or "Dr. Lucy," like they'd been posted to a friend. It's as if, among the predictors, she is, by turns, larger and smaller than life, a mythological figure, a mirror for their fixations, yet at the same time accessible, someone with whom they can build a relationship (imagined though it may be). The more I consider this, the more I can see the connection; the first time I noticed Jones, in the wake of the 1991 Sierra Madre earthquake, I, too, was drawn to her flat, open features, her matter-of-factness, the subtle reassurance of listening to someone who seemed to know what was going on. That's the genesis of my interest in seismology—that very moment, watching Jones on television—which makes the act of reading all this correspondence feel uncomfortably familiar, forcing me to think about the link between prediction and personality, the types of stresses that trigger earthquake forecasts, not just on the seismic level but the individual one. "Look," Jones says about the interface between the USGS and the public, "part of what we're doing is psychology. We're taking something completely random like an earthquake, and giving it a name and number, placing it in definable terms we understand. People need that, to take control, to exert control over an uncontrollable situation; it helps them get a handle on their fear." Setting aside for a moment whether or not earthquakes really are so random, this is what I'm after also, this sense of truth, of larger patterns. The problem, Jones goes on, is that, too often, the need for patterns overwhelms all else, even common sense, which she believes is the one true story told by Curtis's prediction files.

Among the predictors Jones finds most troubling is a man named Donald Dowdy, who, during the late 1980s and early 1990s, sent her a bizarre series of manifestos, postcards, rants, and hand-drawn maps, forecasting full-bore seismic apocalypse around an elusive, if biblical, theme. Dowdy's peculiar pathology was to insist that earthquakes not be interpreted in terms of science but of religion, as expressions of redemption through destruction, evidence of God's refining fire. In a letter dated January 20, 1989, he informed the USGS that "My views of The Big Picture may come from not so Scientific applications as you would prefere [*sic*] to deal with, But, I am understanding, more & more the SPIRITUAL Implications of what IS happening in this City of Angels. . . . 'I HAVE BEEN ACTIVE; AS SUCH AS THE LAND I WALK.' " In a second message, this one undated, he took a somewhat more poetic tack:

I Like The Sounds of—Breaking Glass,
& I Like The Sights of Cars—About to Crash,
& I'd Like to Turn Your Party—Into a Bash.
I Like the Sounds of—A Breaking Heart
& I Like the Sounds of Lovers—Tearing Apart,
& I'd Like to Turn Your Party—Into a Bash

Cause you see me when you Blink & you never
Stop To Think That the Light Inside—may Well,
May Well, May Well, Be Alive!
I Like the Sounds of—Plates Collideing.
I Like the Sights of—Mountians Devideing.
& I'd like to Turn Your City—Into a Bash.

Often, Dowdy's notes arrived written in his own invented language; "Lucy—" reads one scrawled across the interior of a store-bought New Year's card, *"Ansin Tiomna Bheith Aon A'Bhalmhor Crith Talún Cathain An Colúr Tiomna Bheith Agflovain Thart Lastuas Ancathhair De —."* Inscribed in an odd, looping calligraphy that merges script with large block letters, and signed *"Aingeals,"* one of a few recurring noms de plume, Dowdy's New Year's greeting is an eerie dispatch from the obsessive underside of California, an alternate universe bounded by complex fixations, strange enough to laugh at, but disturbing also, as if there's something going on here that you just can't put your finger on.

What distinguishes Dowdy from other predictors—even the handful who, like him, interpret earthquakes as expressions of apocalypse—is his sense of intent, of direct engagement; he sees devastation as a matter of personal cause and effect. That's because, according to his cosmology, Dowdy is not just a forecaster but a seer, and not just a seer but a catalyst, a seismic trigger, operating on a direct pipeline to God. As he proclaims in one long handwritten draft, by way of predicting an 8.7 on the San Andreas that will submerge the entire Los Angeles basin twenty-five feet beneath the Pacific:

> I am a Melchizedek prophet & I am led by the spirit of God to these things, to share with you as my eyes are quickend [*sic*] & open to behold as is shown, to me.

> My authenticity is 100%, & I will to do as is requested by my father God; & tell you, as well as many others, that our Oct. 1st quake was much more than that. It was a *warning*.
>
> I was here Feb. 9th 1971. I lived it then. I was in Mammoth Lakes in 1980 quakes. I stomped my foot on the Andreas rift in Palm Springs in 86 & that area shook awhile, but, I'll tell you, I never was more terrorfied [*sic*] as to the aspects Oct 1st's jolt felt.

Lest there be any confusion about his meaning, Dowdy (or D+Cmerica, as he calls himself in this particular incarnation) goes on, "I don't want to have to bring forth more proof, like a 7.1 on the San Fernando fault, so just listen, & give heed to the warning." The statement echoes another, earlier declaration, from a 1987 postcard: "I do know what I'm doing. I am a walking quake."

Jones won't talk much about Dowdy, nor, for that matter, will anyone else at the USGS. For them, he represents the dark heart of the predictor phenomenon, less a harmless crank than someone who is seriously deranged. Certainly, there's ample justification for this assessment; in March 1993, four years into their correspondence, Dowdy sent Jones a road map of Los Angeles County on which he'd written, "One of these days, I'm going to cut you into little pieces," a reference to the Pink Floyd song "One of These Days." According to the official Caltech Security Incident Report, "This was interpreted as a threat by USGS personnel," and the FBI was called. Two months later, after a visit from federal agents, Dowdy wrote Jones a formal letter of apology, pleading, "I have no intention of harming You, or Anyone at Cal Tec [*sic*], Or Anyone anywhere. The line 'One Of These Days I'm Going To Cut You Into Little Pieces' is The Lyrics from a Pink Floyd song Titled the same. I heard it while in Meditation seeing a vision of a great Quake. It referrs [*sic*] to the Geology of So. Calif.. Being written on a map of the area, showing the fault lines, it means THE BIG ONE." He closed the note by writing, in a tone so plaintive you can almost hear it reverberate off the paper, "Thank You for at least reading My Mail."

There's something heartbreaking about Dowdy's apology, not least because it feels unbearably intimate, like a secret publicly exposed. More discomforting, though, is the letter's wistfulness, its almost palpable yearning, which is so intense it leaves me unsure of my own standing, unsure what to believe. I need to be clear what I mean by this: Without doubt, Dowdy's comments, his fixations, are disturbing, and I don't want to minimize the sense of hazard they evoke. Yet at the same time, I can't help thinking that the trouble here may also have to do with the disconnect between science and popular culture, the chasm between research and the outside world. It might not be surprising that no one at the USGS would pick up a Pink Floyd reference, but for anyone familiar with the song, the lyric jumps out at you immediately. Faced with that, I wonder if the USGS's reaction to Dowdy is a matter of context, of misinterpretation, as much as of security. Is his note a threat, or is it just a seismic warning? Is it targeted at Jones, or is it a reflection on the geology of the region, the incipient disruption that is the only thing about which Jones and Dowdy agree?

Questions like these will drive you crazy if you let them; they are circular questions, questions without answers, questions that slip against each other like earthquakes on their own small, psychic faults. The more I sit here, however, the more I find myself caught up in their strange interiority, the less I'm able to turn away. It's not that I believe Dowdy, or think he's a prophet: how could I, when his theories revolve around the notion that, in the pattern of the L.A. freeway system, there is an apparition of a dove whose presence serves to restrain "the forces of the San Andreas fault"? Still, as I spread his maps and letters—pages and pages of them—across the surface of the conference table, I feel the hint of something, a vague and sneaking sympathy. It's impossible for me to read these notes and fragments without trying to put myself inside the mind of their creator, to imagine what it would be like to experience the world from his perspective, to fear uncertainty so much that I retreated into my own internal hall of mirrors. How does a man like Dowdy live? How does he shop in a grocery store? How does he hold a job, or go out in the evening? How does he fall in love? "I see prediction as having to do with severe emotional issues for a lot of people," Lucy Jones once told me, but despite the accuracy of that statement, it's too abstract. No, I'm after the connections, the strands of value, of invention, that form the fabric of the predictors' universe—the interior processes by which they make sense of their lives.

The knock on California is that we're all like this, which to some extent, I guess, is true. What do *I* want from earthquakes but a mythic structure, a set of meanings that, if not entirely physical, is at least philosophical, intuitive, and whole. The days of William Money, though, are well since past us, and this is no longer a landscape where oddballs and visionaries can carve out their own corners of existence, wholly disconnected from the status quo. Money, writes Carey McWilliams, inhabited "a weird oval structure in San Gabriel, the approaches to which were guarded by two octagonal edifices built of wood and adobe, [and] was the leading Los Angeles eccentric from 1841 until his death in 1880." Dowdy, on the other hand, lives (or once did, at any rate; his dispatches end abruptly with his letter of apology) on an anonymous boulevard somewhere deep in the San Fernando Valley, in a neighborhood that I imagine is much like thousands of others across Los Angeles, full of apartment complexes and nondescript stucco houses, flat and sun-bleached and unchanging for as far as one can see. Money, in other words, could transform his world, his literal surroundings, by the force of his bizarreness, whereas Dowdy . . . Dowdy is adrift in a universe that has long since been configured for him, where the only territory left available for transformation is the one inside his head. Even his driver's license, photocopied and enlarged as part of the USGS complaint, doesn't yield much besides the basic statistics: height and weight and date of birth, and a signature scrawled in the handwriting of an overgrown child. There's a picture, grainy from the copying machine, of a long-haired man with a lantern jaw, eyes set wide within an oval face. I keep thinking that those eyes might tell me something, but the more I look into them, the more impenetrable they seem, black pinpoints on a sheet of paper, so that no matter how hard I stare, they remain closed to me.

About an hour later, I'm back in Linda Curtis's office, sitting in an armchair just inside the doorway, pondering the fluid shape of myth. Whatever their intentions, predictors like Joyce and Dowdy are involved in the elaboration of contemporary legend, although in each case, it's a legend dependent on its own eccentricity, one that looks inwards, towards the muddled private landscape of the individual, rather than outwards, towards the world. This, Lucy Jones would tell you, is the hallmark of nearly everyone who has ever staked a claim to quake prediction; "The truth," she says, "is that we deal with hundreds of these people, and most of them get obsessed with their own theories. They don't want to have to measure their ideas against the same standards that I do, which contributes to a lot of misinformation about how earthquakes work." But while in some sense, Jones is clearly right about that, I can't help wondering if there's more to the story, another angle, a midpoint at which intuition and delusion coincide. It wouldn't be the first time; it may be impossible, these days, to imagine geology, or earthquakes, existing outside a system of plates and faults, yet when tectonic theory was first introduced in the 1960s, most scientists reacted with ridicule and disbelief. If the entire structure of geology could shift so radically with a single concept, why can't it change again, in equally dramatic fashion, making science out of areas we currently define as myth? Either way, you have to admire the fortitude of someone who sticks to a hypothesis in the face of utter rejection—by the scientific community, by the public, even by the earth itself. It gives you an idea of just how powerfully belief can sustain us, which is, of course, a necessary aspect of living in an earthquake zone.

For Curtis, the whole issue of belief asserts itself most vividly in the area of earthquake sensitivity, a subset of prediction for which she feels a particular sympathy. Or not sympathy, she corrects me when I use that word, but *em*pathy, identification, even personal appeal. "I know you're going to think I'm crazy," she laughs, ducking her head again as if for self-protection, "but the sensitives are of interest to me because I have migraines, and I sort of relate to what they're saying. The morning of the Seattle earthquake, I had to go to my doctor, I was so severely ill. I was throwing up from a migraine, and I had to go to the doctor and get an injection. It was totally out of the time frame when I would normally get a migraine. So I was like . . . hmm." Listening to her, I'm aware of an instant when the conversation starts to break wide open, as if we're on the cusp of something, as if, for a second anyway, she's about to cross the line that separates her from the X-Files and its denizens. Just as quickly, though, the moment passes, and she backs off a little, telling me that migraines, just like earthquakes, tend to come on randomly, writing off the whole thing as coincidence. Still, Curtis insists, there's something different about sensitives, about the way their predictions arise from physical symptoms, something less conflicted, more genuine. "I can envision when it's going to rain," she says. "I have two broken fingers, and I broke my wrist roller skating, and when it's going to rain, I can tell. So I have a sense that people who are sensitive to earthquakes are sensing something similar."

Curtis's analogy is so down to earth, so recognizable, that I find myself nodding in agreement—and not just because it sounds like common sense. I, too, have a pair of broken fingers, and ever since I injured them, when I was ten, I've felt the joints swell and stiffen before rainstorms like an internal barometer. Why, then, shouldn't earthquake sensitivity work like this, not so much a matter of intuition as of physicality, a way of tapping into the movements of the earth? Certainly, if you listen to the sensitives, many make a case for just this type of connection, in which an impending quake announces itself as a sensation: a noise, a body ache, or a discomfort, a pressure that needs to be released. One such predictor, a fifty-five-year-old electronics technician at the University of California, Irvine, named Jack Lockhart, has even written an essay called "Symptoms of a Sensitive," which first appeared a decade or so ago in the publication *Geo-Monitor*, and has had something of a second life on the Internet. In it, Lockhart methodically details the terms of his sensitivity, identifying a wide array of sounds and smells (TURN & BURN: "Ever smell aluminum or sheet metal being cut by a saw; an electrical cord being burned; a battery venting—giving off fumes or vapors? That is exactly what I smell here"; FRUIT LOOPS: "Mix the juices of a watermelon, grapefruit, lemon, melon and orange together and you get an idea of this smell. It smells real sweet. This smell scares me because it represents an event that will injure or take someone's life"), as well as headaches, nausea, diarrhea, and something he calls "q-flu," or quake flu, in which flulike symptoms rack his body, only to dissipate once a temblor strikes. "Do I believe in myself?" Lockhart asks at the end of his essay. "When I first recognized my 'gifts' I was very skeptical. NOW; You can bet that I wouldn't play on the rim of the Grand Canyon if I were vacationing in Arizona and experienced 'turn and burn,' 'fruit loops,' 'q-flu' or 'bump & grind.' "

In much the same way as Lockhart, there's a lot about sensitivity that leaves me skeptical, a lot of crazy correlations, predictions that don't pan out. I don't know what to make, for instance, of the Barstow, California, man who writes that, prior to an earthquake, both he and his wife experience heaviness in their legs so profound they can't get out of bed, nor of the predictor Curtis has nicknamed "Pain-in-the-Butt Man," because he feels pain shoot through his ass cheeks before the ground begins to shake. At the same time, I'm oddly drawn to someone like Lockhart, who no longer issues public predictions because he's tired of being ridiculed, or Kathy Gori, with her phenomenal record of success. As strange as it may sound, these people seem stable to me, credible: they don't make outrageous claims. Gori is a screenwriter and former morning news radio coanchor; she is smart and cynical about the nature of belief. "I come to this," she says, "in purely pragmatic fashion. I'm not a lunatic. I don't wear tinfoil underwear. It's a natural thing." Her predictions reflect this matter-of-factness; if an earthquake is too small or too far away (anywhere outside Southern California), she can't feel it, a limitation that, to my mind, boosts her credibility. Lockhart operates within similar geographic boundaries. Except for the 8.1 earthquake that struck Mexico City on September 19, 1985—"I got real sick and real sad," he recalls. "I knew there was going to be an earthquake, and a lot of people were going to die"—his sensitivity peters out past a distance of five hundred miles. The more I think about this, the more I wonder the extent to which sensitives like Gori and Lockhart are picking up on a subterranean shorthand, a set of not-yet-understood connections with the earth. Is something really going on here? How much should I believe?

Perhaps no one embodies the push and pull of these particular questions as much as Charlotte King, the self-styled doyenne of the earthquake sensitives, a fifty-five-year-old Salem, Oregon, woman who, over the last two decades, has inundated Linda Curtis with so many predictions that she has her own subsection of the X-Files, a blue folder full of e-mails, press releases, and biographical information that raise as many issues as they resolve. Since 1976, King claims, she has literally been able to *hear* low-frequency sound waves—a foghornlike moaning she refers to simply as "The Sound"—and, in conjunction with physical symptoms ranging from anxiety and irritability to nosebleeds, muscle spasms, headaches, and severe stomach or heart pain, use them to predict earthquakes and volcanoes with a rate of accuracy that, by her accounting, comes in somewhere around 85 percent. Such a figure, obviously, is open to interpretation, yet even by the most stringent standards, King has nailed her share of predictions, including the eruption of Mount St. Helens, and the Whittier Narrows, Loma Prieta, and Landers earthquakes; her call to the USGS on the Thursday before Northridge, in which she reported sensing something big in Southern California, is one of the few forecasts Linda Curtis says still lingers in her mind. King has also found what she considers to be direct correlations between her sensitivity and those of animals, particularly whales, who, she believes, hear the same things she does and become confused, to the point of beaching themselves, "trying to escape the low-frequency sounds." There's a certain folk logic to such an argument, for animal sensitivity has long been a component of earthquake lore; history, not to mention Curtis's file folders, features countless reports of horses stampeding, cats and birds panicking, and dogs running away in the days before a quake. As early as 373 BC, writes Helmut Tributsch in *When the Snakes Awake: Animals and Earthquake Prediction*, just before the ancient Greek city of Helice was destroyed "by an especially violent earthquake, . . . all animals that had been in it, such as rats, snakes, weasels, centipedes, worms, and beetles, migrated in droves along the connecting road toward the city of Koria." It was King who first recommended Tributsch's book to me, and reading it gives me an idea of how she sees herself, as part of a seismic heritage, a continuum that extends far beyond the edges of a single life. The same, of course, could be said for geologic time, and recognizing this gives me a little jolt of understanding, as if, in that impulse also, the worlds of seismology and sensitivity may not be so distinct, after all.

Yet just as I feel I might have a grasp on something, I'm reminded of how thin the membrane is between impulse and obsession, an obsession that comes to define King the more familiar with her you become. It's not that I have trouble believing in her sensitivity to low-end electromagnetic frequencies ("I'm a person who responds to changes in the earth's electromagnetic field," she says), or that this, in turn, might enable her to "hear" the subtle shifting of seismic plates about to slip. But unlike Lockhart or Gori, King can't stop herself from stretching the boundaries, especially when it comes to the all-encompassing nature of her predictions, the way they're not restricted by geography or distance, while her ailments . . . let's just say that each area of her body corresponds to a specific region of the world. Her daily e-mail updates lay out so many symptoms and forecasts that they bleed into the ridiculous. "Most of the pain today has been in the sturnum [*sic*] again as well as the upper back," reads one typical installment,

> mostly on the left side, (JAPAN) had to take muscle relaxers as soon as I got up, thought I had lifted something wrong, but did not lift anything.
>
> Upper back as I have repeatedly said is for Japan, left side, that would be for(L) E/S Honshu, Hokkaido, and Kyushu/Tokyo (R) is Kuril Islands entire upper back Indonesia including Java, Jakarta, Irain Jawa, Suwesli Indonesia.. and Halmahera Indonesia.. and others.
>
> Had several hours of right side chest pain again, last time that happened it was Mt. Etna. so I am watching Etna, Kilauea and Pinatubo, and Mexico's Colima volcano.. the Alaskan volcanoes also cause this symptom. Heart is also Mexico, coastal areas including Jalisco, Guerreo, Colima, and Chiapas . . .
>
> Pain in ears, this is usually Italy, Sicily, Greece and Crete . . . with heart pain it is usually Italy and Central Greece.. as well as some Southern Ca. locations.
>
> Low back is still sore, so I am still waiting for one of the Wasatch/New Madrid locations I always list to hit again.. most probable are Utah, Montana, Wyoming, and Idaho/Eastern Idaho . . . then it goes into Arkansas, Alabama, Tennessee, and Missouri, PA. with about 14 more states being along the line of affect.

Were even a fraction of these predictions to pan out, we'd be living in a fluid universe, where every day brought seismic changes on a catastrophic scale. And to be fair, it's true that, on some level, every day *does* bring earthquakes, dozens, hundreds of tiny tremors, as if the earth were an old house, creaking as it settles in its beams and joints. There's a difference, though, between these small, unnoticeable spasms and the kind of ground-shaking, life-changing disruptions King claims to announce. "I try to ignore most quakes under five," she says. "I don't consider them valid predictions because they're too small." Yet if such a standard makes King seem rigorous, it is here, ironically, that her record, her 85 percent accuracy rate, begins to break down. "It's a shotgun approach to earthquake prediction," Linda Curtis notes, with just a trace of irritation. "You scatter your predictions far and wide, and you're bound to have some of them stick." Curtis has a point; if you read through a month or two of King's e-mail updates, you come away with the inescapable feeling that she is hedging her bets. At the bottom of each transmission is a list of the previous day's earthquakes, many of which she links retroactively to a prediction, or some physical distress. Especially problematic is her propensity to predict aftershocks, which even the most skeptical seismologist will tell you can be anticipated (if not forecast outright) with a pretty good degree of success.

In some ways, this brings us back to the conundrum marked out by every earthquake sensitive, each time he or she asks us to measure the relationship between, say, a bout of vertigo and a seismic event hundreds, even thousands, of miles away. With King, however, these issues only grow more complicated as prediction yields to an almost Dowdy-esque sense of fantasy, until what she's describing seems less a system for interpreting earthquakes than an all-encompassing personal mythology, an iconography of empowerment, with even the most random events interpreted through the fulcrum of her "gift." Although King is adamant that she's "not a psychic," if you let her talk long enough, the conversation veers in inexplicable directions; she makes connections between certain earthquakes and apparently unrelated phenomena such as train crashes or acts of random violence, and argues, with a quiet fervor, that babies who die of crib death may, in fact, be infant sensitives, overwhelmed by low-end electromagnetic stimuli until they literally forget to breathe. When I mention this to Linda Curtis, she grins and rolls her eyes a little, as if this is something that she's heard before. "That's the whole thing with Charlotte," she laughs. "She's so far afield with all this stuff. For me, the breaking point came when she started talking about alien abduction. I wanted to focus on earthquakes, mainly in California, Oregon, and Washington, but she was all over the Wasatch, and New Madrid, and telling me about plane crashes and domestic violence."

"I know," I say, remembering one late night phone conversation when, in a tone so calm and uninflected she might have been discussing the weather, King insisted that "any time someone goes home and kills their family or there's a murder-suicide, anything spur-of-the-moment or really violent, you can be sure that within one to four days will follow a quake in the Chile-Bolivia-Argentina border area." For weeks, I tried to wrap my mind around the concept. What does it take, I wondered, to accept this? Is it just a fear of randomness, or does it speak to something larger, some need to tie up all the loose ends, to smooth the rougher edges of the world? The temptation is to write it all off to psychology; as a result of her symptoms, King has long held legally disabled status, and her problems (or so she tells me) helped undermine an eighteen-year marriage and relationships with three children she was too sick to care for most of the time. But as compelling as I find that argument, I can't help feeling, underneath, a deeper reason, a note of . . . *sensitivity*. Like other sensitives—Jack Lockhart, for one, who describes the souls of those who will be killed in major earthquakes: "You've seen birds flying?" he asks rhetorically. "I see their souls leaving their bodies, leaving this earth, going to wherever it is they go"—King knows well the burden of belief. "Not to appoint myself their savior," she says, "but it took a long time after those ten thousand people died in the Mexico City earthquake that I didn't feel responsible." There's something so genuine about those sentiments, so sad and moving and sweet and human, that it makes me want to overlook the inconsistencies, the delusions, and offer up the benefit of the doubt. Or maybe it's just that, as I admit to Linda Curtis, "A lot of the time, when I talk to people like Charlotte, they start off sounding pretty reasonable . . ."

"Yeah," Curtis interjects. "They all start off sounding reasonable. Until they go too far."

Towards the end of the afternoon, just as I'm about to pack up my notes and information, something happens that reminds me once again of all the odd synchronicity that attends earthquakes, the signs and symbols, the coincidences, the geopoetry. Curtis has been telling me about her experience taping the TV show *In Search Of*, which she's just done with a predictor named Zhonghao Shou. Like Charlotte King, Shou—or Cloud Man, as he's known with some affection around the USGS office—is responsible for a significant chunk of the material in the X-Files; over the last decade or so, he's issued hundreds of seismic forecasts, which he makes by studying the clouds. "They were doing different segments," Curtis explains, referring to *In Search Of*'s producers, "and Cloud Man was one. They did a touchy-feely thing where they had me invite him in, and let him park his bike inside. They wanted me to look at some of his predictions, but because we'd been talking about the star stickers I put on the ones where he gets five or six elements right, they didn't want to see any that didn't have stars. They wanted the stars there, and to see me leaf through those. It was all very staged."

On the one hand, Curtis's story further illustrates the difficulty of getting any kind of definitive read on prediction, with a program like *In Search Of* literally skewing the data to make a better broadcast, ignoring Shou's mistakes or misinterpretations even as it sensationalizes his hits. But more than anything, it makes me feel bad, complicit even, as if, just by listening, I'm helping take advantage of Shou myself. In many ways, after all, Shou is the one X-Files predictor who *doesn't* go too far, a sixty-three-year-old Chinese former chemist who is the very image of quiet dignity. Sometimes, walking around the Caltech campus, you can see him, riding his bicycle along the streets and pathways, or talking with his daughter Wenying, a PhD candidate in biology who acts as his translator-advocate, accompanying him to interviews and deciphering his stilted English. Except for the almost incomprehensible thickness of his accent, Shou could easily pass for a professor; neatly dressed, with thinning hair and a long, sleepy face, he looks like he belongs. The same is true of his predictions, which always come precisely typed, or written out in a calligrapher's hand. Featuring graphs, charts, diagrams—even statistical analyses that frame his forecasts in terms of probabilities—they rely on their own hybrid form of scientific jargon, which, whether or not it lends them any additional credibility, does provide a vivid insight into Shou's determination to be viewed not as some fringe enthusiast but as a legitimate researcher, seeking purchase on a landscape of which Caltech is the absolute epicenter of the world.

Then, no more than five minutes after Curtis has finished talking, there is a rustle in the hall outside her office, and the front door of the building swishes open and shut. A moment later, Shou himself appears in the office entry, as if her story had invoked him, like an incantation or a spell. Freshly pressed as usual in ironed blue jeans and black loafers, he pushes his bike ahead of him, fingering a shoulder bag. Curtis shoots me a quick look—of bewilderment, or wonder—as Shou says hello in a low voice, bowing his head slightly and giving her a toothy grin. Briefly, his eyes pass over me, still sitting in the armchair, but it's been a long time since we've seen each other, and he doesn't recognize me. It's just as well, really, for we're not on the firmest footing; after I first wrote about him, a couple of years ago, Shou posted a long letter on his Web site accusing me of misinterpreting his ideas, with the intent of ridiculing him. He's also mad at me for having lost a small envelope of his cloud photos, including one, taken in Pasadena in early January 1994, which, he claims, presaged the Northridge quake.

Shou leans his bicycle against the side wall of the office, and begins to dig around inside his bag. In the meantime, Curtis rises from behind her desk, while I try to disappear into my chair. "How are you?" she asks, approaching him, as the late afternoon sun slants across the floor, cut by the venetian blinds into alternating ribbons of shadow and light. It's a thick light, grainy with motes of dust, and it gives the room an air of indistinction, as if the boundaries of the world have been blurred. If anything, in fact, the layered lucency makes Curtis's office look like a stage set, as if what I'm seeing is that *In Search Of* segment come to life. But while it's a little eerie, all this confluence, I can also tell there's more at work—that, without the cameras and the staging, some kind of genuine interaction is going on. It doesn't take much, after all, to see that Shou and Curtis like each other, that, if circumstances were different, they might actually be friends. As I watch, Shou fishes a magazine out of his satchel and folds it open; "Oh, a picture and everything," Curtis exclaims after he passes it over, and when he indicates that she's to keep it, she adds, "Good. I'll put it in your file." They talk quietly for a few minutes; then, Curtis asks if he's seen any earthquake clouds lately, and he answers with an animated yes. "I thought I saw one the other night," she says, "but my husband said it was a jet trail." She laughs, but it's a gentle laugh, not malicious, which only reinforces my sense that there's a bond here, whatever Curtis may think about Shou's prediction model aside.

Of course, the thing about Shou's prediction model is that Curtis isn't sure what she thinks, exactly; "When we were on *In Search Of*," she admits, "they asked if I understood how his predictions worked, and I told them, 'The only thing I know about Cloud Man's predictions is that he says they're giant finger-pointing clouds. They point like a finger down to where your earthquake's going to be.' " That's a bit of an oversimplification, but not by much, for if Shou's theory has a hallmark, it's how absolutely uncomplicated the whole thing seems. The idea is that, as stresses build throughout a fault zone prior to an earthquake, the subterranean rock will start to crack, leaving space for water to collect. When enough pressure accumulates, water levels—and, more important, temperatures—go up, producing vapor that gets pushed to the surface of the fault. "Through a gap," Shou explains in an unpublished manuscript, "Earthquake Clouds and Short Term Prediction," "the vapor rises up and floats following the surface wind. Meeting the cold air, it forms a cloud. The shape of the gap and surface current may endow the cloud with a special configuration like a snake, a wave, a feather, or a lantern etc., which will be able to be distinguished from weather clouds."

The belief that clouds can be useful in predicting earthquakes is hardly a new one; like sensitivity, it's been around for thousands of years. In *When the Snakes Awake,* Helmut Tributsch traces it all the way back to ancient Egypt, and quotes Aristotle and Pliny the Elder on the subject. "When an earthquake is impending," writes the latter in his *Historia naturalis,* "either in the daytime or a little after sunset, in fine weather, it is preceded by a thin streak of cloud stretching over a wide space." Shou himself cites yet another example: "There was a document in the Lon-De County Chronicle, China, 300 years ago (recompiled in 1935): 'It was sunny and warm; the sky was blue and clear. Suddenly, there appeared threads of black clouds spanning the sky like a long snake. The clouds stayed for a long time, so there would be an earthquake.' " Among the Japanese, the phenomenon is so well known that there's even a word for it—*chiki*, or "air from the earth." Just before the 1805 temblor that rattled the Japan Sea island of Sado, an engineer evacuated a mine there after noticing an unusually thick ground fog, and if you look at the official report on the disaster, notes Christopher Scholz, a professor of tectonophysics at Columbia University's Lamont-Doherty Earth Observatory, his observations are treated as "completely factual. There was never any question about what he meant."

For Shou, the existence of such a long-standing documentary record is one of the most important arguments in favor of cloud prediction, the thing that sets his system apart. How can you argue, after all, with Aristotle or Pliny? How can you dismiss so much history? But Aristotle, let's not forget, is the philosopher responsible for the wind-in-underground-caverns theory of earthquakes, while Pliny, for all his attention to precursors and forecasts, was killed in the eruption of Mount Vesuvius. And when it comes to an idea like *chiki*—whether or not it represented state-of-the-art earthquake science in early nineteenth-century Japan, that was nearly two hundred years ago. To the skeptics, it's the seismic equivalent of looking at the human body through the filter of Robert Burton's "humours," or bleeding patients with leeches when they're sick. "Earthquakes," scoffs Lucy Jones, "happen ten kilometers below the surface. You can't create a cloud down there." Former USGS seismologist Jim Mori agrees: "If it were so obvious, it should be obvious. A lot of temperature changes would have to take place at ground level to make large clouds, and we haven't seen them. There's quite a bit of monitoring going on, and it's hard to believe we're missing all that stuff."

And yet, and yet, and yet . . . I can't help thinking that, more than anyone in the X-Files, Shou does seem to be approaching, even incrementally, some strange shadow of the credibility he seeks. Of all Curtis's predictors, he's the only one who retracts forecasts, and in an essay posted on his Web site, he suggests that people judge him by, among other things, looking at his failed predictions, to get a better sense of what's at work. Even his successes tend to be uncommon, if only because so many of them are off the beaten track. For the majority of predictors, the issue's not just how but *what* you forecast; there's a lot of emphasis on hitting the more visible quakes, like Loma Prieta or Taiwan or Kobe, the quakes that get everybody talking, that remind us, with a visceral immediacy, just how fleeting is our sense of solid ground. Shou, however, doesn't seem to care about that one way or the other. Although he's called his share of notable events—among them the 6.8 Seattle earthquake that made Linda Curtis feel so sick (the night before, he predicted that a temblor of magnitude five or higher would soon hit "Canada and Washington State")—he's just as likely to predict a tremor in the desert somewhere, or in a remote region of, say, Iran. In mid-August 1999, he even refuted the NASA Jet Propulsion Laboratory's warning that the next major Southern California earthquake would probably take place in "the heart of" Los Angeles, arguing instead that the real risk lay further east. Two months later, in a desolate corner of the eastern Mojave, the 7.0 Hector Mine earthquake stunned the seismological establishment when it struck the Lavic Lake Fault, a fault previously declared inactive by the California Division of Mines and Geology, since there had been no large earthquakes on it for more than ten thousand years.

Where such a moment leaves me is with a shock of possibility, the sense that I am standing at a precipice, an escarpment separating what we can imagine from what we might actually come to know. It seems I'm not alone in thinking this; since 1994, Shou has twice been called in to meet USGS geologists, most recently just a few months ago. As with most aspects of earthquake prediction, though, it is here that answers become elusive, become, Curtis notes, "subjective just like anything else." In Shou's case, confirmation is made more difficult by the range of his probabilities; his forecasts, says Jim Mori, are "pretty large in terms of the area and times they cover," incorporating an array of potential longitudes, latitudes, and magnitudes, and a time span of up to forty-nine days. It's not that earthquakes don't occur to match his numbers, just that no one can be sure if the predictions fit the tremors, or only *seem* to because the parameters are so wide. This, in turn, speaks to a larger difficulty with assessing earthquake forecasts, since, as Thurston Clarke argues in his book *California Fault*, "California was so seismically active, any prediction relying on a 'seismic window' of several weeks hedged with a 'probability of 70 to 80 percent' was a reasonable gamble." Either way, preliminary findings indicate that Shou "gets about as much right as anyone could from guessing, maybe just slightly above," and at one point, Linda Curtis tells me, there was a stretch where he had nineteen misses in a row.

If you look at those nineteen misses one way, they seem a lot like nineteen cloudless holes in Shou's prediction theory, nineteen question marks. Yet view them through a slightly different filter, and you're once again dealing with the transformative power of belief. Misses or no misses, after all, Shou is here right now in Curtis's office, talking about earthquakes, as he is once or twice a week, *every* week. To be sure, there are practical reasons for this; by his own count, Shou has spent thirty thousand dollars on research since the early 1990s, and, like Charlotte King, he's cut himself off from a significant part of his past, leaving a wife and a career back in China—which means he's got a lot at stake. Yet watching him, I see no real sign of pressure, no particular fear of falling on his face. Rather, he appears to take a certain pleasure, even sustenance, from these challenges, from the ongoing efforts to validate his work. When Curtis asks if he has a new prediction, Shou says yes, withdrawing a small map and pointing out the corresponding cloud patterns, suggesting possible epicenters and dates. After a minute, he and Curtis move to the copy machine and make a duplicate of his forecast, heads bobbing together almost conspiratorially, silhouetted like a pair of cameos in the musty end-of-afternoon light. Once they're done, they slip back over to Curtis's desk, where they call up a shake map of California on the computer, as if, in the interstices between light and pixels, they might actually be able to see Shou's earthquake as it comes to pass.

"Is it near China Lake or farther north?" Curtis asks, clicking her mouse to enlarge the image on the screen.

"Near China Lake," Shou answers, adding that he's not sure about the magnitude, which could be either in the five or seven range.

"Let's split the difference and make it a six," Curtis offers, and he agrees with a laugh. Then, Curtis signs and dates both copies of his prediction, and, with a certain quiet sense of ceremony, hands one back to Shou.

Just as Shou and Curtis finish, I decide to bite the bullet, and lean forward to introduce myself. When Shou hears my name, he nods once, sharply, and his mouth tightens into a thin white line. I extend my hand, and he shakes it, but no sooner are we finished than he asks about the photos I've lost. I may have trouble with his English, but I have no problem understanding what he's saying, which is that he still holds me responsible for those pictures, and he wants them back.

"I'm sorry," I murmur. "I haven't found them. But I can look again."

"I need the pictures," he says. "I don't have other copies of those clouds."

"I know, I know," I tell him, feeling bad and slightly bothered all at once. "But in the meantime, listen; I've been meaning to call you—I want to talk to you about cloud prediction again."

"I need the pictures," Shou insists again.

"I'll look for them," I say. "But are you willing to talk to me?"

There is an extended pause, just long enough for my chest to prickle with discomfort, and my eyes to grow restless peering at his face. Shou is looking at me as if he's trying to decipher something, as if he thinks I've got those pictures hidden away. The truth is I could probably find them, although I really don't know where they are. But even if I wanted to tell Shou this, I'm not sure I could find the language, and anyway, before I have the opportunity, he closes down the conversation, barking at me: "Get the pictures. Then, maybe, we can talk."

On my way home from Pasadena, I take the 110 freeway into downtown Los Angeles, staring up at the tall buildings as the road slices a submerged path through the city, which rises alongside it like an illusion or a mirage. When I first moved here, I used to refer to downtown as the Emerald City, since it was always shimmering somewhere in the distance, a destination I could never get to, a center I could not hold. Now, I think of it that way for another reason, which has to do with my sense of all these buildings as less than solid somehow, as dreams that could dissolve in the flash of a fault line slipping, tumbling down to nothing, tumbling down. There are faults in downtown—one, discovered in 1999, was determined, retroactively, to have triggered the 1987 Whittier Narrows quake—faults that run along Wilshire and Sunset boulevards, faults we do not know about, faults we cannot see. There are, Mike Davis warns in *Ecology of Fear: Los Angeles and the Imagination of Disaster*, "more than 50 active faults directly underneath the heavily urbanized portions of Los Angeles, Orange, and Ventura Counties"—a fact that renders the entire Southern California megalopolis "at least in theory, 'epicentral' to a massively destructive quake." In the face of that, I think, slipping south of downtown, it's no surprise that we look for certainty wherever it suggests itself, that what we know for sure often feels less essential than what we believe. As I take the turnoff for the 10, and the city disappears in my rearview mirror, I can't help but wonder how long all this will remain there, or even if, in some weird way, it's really there at all.

This sense of unreality, of suspension, ebbs and flows throughout the afternoon and evening, as I have dinner with my family and put my kids to bed. On the one hand, all the small, domestic activity helps to root me, but at the same time, I can't quite shake the feeling that I'm living on the crust of history, that there's a lot below the surface I don't know. Finally, I end up at the computer, where I find my way to Shou's Web site, although it's not theories or predictions I want tonight. Instead, I scroll down to the link that opens up the guest book, an online log where cybervisitors are invited to sign in and give reactions to Shou's work. There are thirty-three pages of entries, featuring more than 130 messages, from both fellow predictors, looking for community, and people living in earthquake country who are eager for information, whatever it may be. I'm not sure what *I'm* after, exactly, but the more I lurk, the more I'm struck by yet another set of resonances, the way these comments echo, without planning or intention, virtually the entire range of prediction theory, from Harvey Rice to John J. Joyce. One man argues that, because "clouds usually carry with them electric charges . . . When the crust of earth moves against one another, electromagnetic radiation will be generated due to the immense friction, and this will affect clouds, and they will form into REGULAR PATTERNS." Another promotes his own prediction system: "A device that defies physics and reveals a very interesting ene energy that migrates in all directions but predominately ene. . . . Its motions relate to meteorology, quakes, animal behavior, etc. etc. It costs less than five dollars to build, or less. prediction is NOOOOOOO problem. maybe i am crazy but maybe i am not."

Reading this, I'm reminded again why so many people see prediction as a sham, a parlor trick, a form of self-aggrandizement or delusion, or, maybe, both. Still, every now and then, I come across a different kind of entry, one striated in fear or longing, one that hearkens back to our desire, our need, to know. "Wow amazing stuff," writes a woman from British Columbia the day of the Seattle earthquake. "We have just experienced an earthquake just south of us in Wash. State and I see you predicted one for approx. feb. 26th. You are amazing. Is Van. B.C. due for one soon?" On another page, a Japanese correspondent begs, "Please advise me if there might be a large Earthquake to hit the Tokyo areas in Japan, so I can prepare myself to a safer grounds. thank you very much, Earth God!" You can virtually feel the panic bubbling like magma deep within those sentences, the unbearable anticipation, the dread and hope all mixed together until it's not clear any longer where each begins and ends. I know that feeling, have lived with that feeling; for the two and a half years preceding Northridge, I spent part of every day trying to imagine what it would be like when the earthquake came, how my house would rock and tumble, how the fabric of the world would be undone. The night the earthquake did arrive, it was almost a relief; I recall crouching against the door frame, clutching my wife, Rae, in a backwards bear hug, feeling shock, glee, terror, utter epiphanic amazement that we'd survived. For all the inconsistencies, the fantasies, the false alarms, this is what prediction offers also, a psychic strategy that, by allowing us for a moment to mitigate our uncertainty, provides an emotional barrier against the impermanence of life along the fault.

Of course, the problem with such a barrier is that it tends to reassure us falsely—or, if you think about it another way, not to reassure us in the least. Even the most acute forecast, after all, cannot prevent an earthquake, but only warn us that a quake is coming, which shifts the burden of conviction from the predictors to ourselves. This feeling, too, is one I know firsthand, for seven years ago, when she was pregnant with Noah, Rae developed a predictive sensitivity of her own. Over the course of a few months, she called three earthquakes and hit on all of them, a small sample but—for me, anyway—a potent one. The way Rae describes it, she would literally *feel* the earth moving, as if a tremor had just rocked the ground beneath her. Then, before the day was out, the sensation would be followed by an actual temblor, after which her sensitivity would settle down. The first time it happened, I shrugged it off as coincidence; the second, I reacted with a fascinated skepticism, as if what we were experiencing couldn't possibly be real. By the third prediction, however, I had grown to trust Rae's instincts, and I rode out the interval between premonition and payoff in an ecstasy of tension, much like the people who e-mail Zhonghao Shou. For the first time, I had to face the key conundrum of prediction: what do you do if you believe? It was like my entire life had been cast into suspension, a situation complicated by Rae's inability to forecast epicenter or magnitude. One hour, two hours, I waited, then another, and another, and another more. By the time the quake rolled in (it was small, 3.2 or 3.4, I can't remember, and centered far enough away that, when its shock waves reached us, the force was no more substantial than a strong breeze), I was on pins and needles, expectancy so great I had to walk the floors. What was the value of prediction, I began to wonder, if you were powerless to stop it? Why, exactly, would you want to know?

That's a question I've never fully come to terms with, and even now I don't know what I think. Yet sitting at my desk, scrolling through Shou's electronic guest book, I can't help coming back to the idea that there's an order in here somewhere, a pattern we can hold. For all the anxiety provoked by Rae's forecasts, they left me with a subtle whisper of connection, as if I had stumbled on the border between fact and folklore—as if an earthquake myth had come to life. These were elemental moments, moments you don't come up against very often, moments that make you feel alive. Thinking about that, I begin to wonder if *this* isn't the fundamental story of the X-Files also, the way that, like all legends, they offer us a way of giving the incomprehensible a human face. It's a contradictory notion, but then so is living in a seismic zone, where the simplest bits of business become extraordinary acts of faith. In such a landscape, the only thing we know for sure is that the earth is moving, which makes belief its own amorphous patch of solid ground. We put our roots down where we find them; we look for meaning where we can.

A BRIEF HISTORY OF SEISMOLOGY

One morning, not more than a month or so after my encounter with Cloud Man, I find myself back at the USGS in Pasadena, visiting the upstairs office of Susan Hough. Hough is one of the two Survey seismologists to have invited Cloud Man in for a meeting, and it's for this reason that I've come to see her, although as keeps happening with earthquakes, our conversation quickly starts to ripple off in unexpected directions like a ground wave propagating out from a ruptured fault. The day is flat and hot, the sky cloudless, seared white with sunlight, and on the other side of Hough's window, workers hammer and saw in stop-and-start sputters, sound erupting when I least expect it, banging out a ragged counterpoint to our words.

Hough has been at the USGS since 1992, following a four-year stint in New York, at the Lamont-Doherty Earth Observatory, where she went to work after finishing graduate school in 1988. Like most seismologists I encounter, she has a healthy outdoor quality about her, open-faced and sturdy, the kind of person you can easily imagine walking rough terrain or hanging out by a tent at night, drinking coffee and swapping stories, waiting for an earthquake to arrive. This morning, she sits at her desk dressed in hiking boots, cargo shorts, and a T-shirt, brown frizzy hair loose and off her shoulders, and when I ask a question, she answers with no artifice, no apparent agenda, other than the engagement of someone who is interested in something that, even now, she can't quite explain. There's an irony to this, if only because of the belief in predictor circles that earthquake science is little more than a vast conspiracy, that geologists spend the bulk of their energy keeping information from the rest of us, *that they really know what's going on*. In July 1992, barely two weeks after a 7.3 earthquake shook the town of Landers, California, north of Palm Springs, the Southern California Seismographic Network, a cooperative venture of the USGS and Caltech, issued "an incomplete list of rumors that we have been receiving literally dozens of calls about today"; among the stories cited were reports that "Caltech knows when the Big One will occur and we aren't telling anyone," and that "a group of seismologists have left Caltech because of the pending quake." When I mention this to Hough, she laughs and spreads her arms in a "Who, me?" gesture, as if to ask what she could possibly have to hide.

What makes Hough interesting is that, of all the seismologists at the Pasadena field office, she's the one most open to the value of folklore, to the notion that there may be something in all those centuries of myths and stories to shed light on the way earthquakes work. For her, it's a matter of professional necessity, since she's spent the last several years studying a sequence of three large earthquakes that struck near the town of New Madrid, in the Missouri Territory, between December 16, 1811, and February 7, 1812. The New Madrid earthquakes, as they've come to be known, represent a fascinating episode in the seismic history of North America, both because of how they challenge our preconceptions about geology and geography (an earthquake in Missouri? thousands of miles from the nearest plate boundary?), and because, occurring as they did in a remote region during a relatively primitive era, they don't provide a lot of the hard data upon which contemporary scientists rely. Instead, Hough's research into these earthquakes has led her into less well-traveled areas of inquiry like historiography and personal history, forcing her to examine anecdotal evidence and compare it to the existing pieces of the geologic record, as a strategy for interpreting what it is, exactly, that went on.

At the most basic level, the idea of using anecdotal information to analyze an earthquake seems to be nothing if not common sense. What, after all, could be more useful than an eyewitness account? Even now, in our highly instrumented, technologically driven era, anecdote plays a role in determining a tremor's size and scope; on its Southern California Web site, the USGS asks visitors to report where they were and what they felt during a given temblor, then uses the information to construct a Community Internet Intensity Map (CIIM)—an electronic shake map that color codes intensity and damage by neighborhood. Within modern seismological circles, however, the CIIM program is less the rule than the exception, a rare point of intersection between narrative and data, between the gray, subjective fear and trembling of an earthquake and the researcher's insistence on framing that in black and white. "There's a bias," Hough admits, "among some seismologists that you can't trust anecdotal material because people exaggerate. And you do have to be careful. Sociologists have done studies of the reliability of oral traditions, in which they take an event that happened, say, fifty years ago, and look at an oral tradition to see how it diverges from the historical record. It's not like you can interpret all of it literally." Still, if anecdote makes for something of a slippery yardstick, it can also yield hints, clues, unexpected validation, information that more empirical perspectives often overlook. "In New Madrid," Hough notes, "the great nineteenth-century geologist Charles Lyall wrote that the native tribes had oral traditions of big earthquakes that predated the 1811–12 events. There's now evidence to support this; we've found traces of two earlier earthquakes, one from around 1415, and another from around AD 900. But for a long time, up until the mid-1980s, seismologists didn't believe that. We thought nothing had happened in New Madrid for ten thousand years."

Hough's comment serves as a vivid reminder of the vagaries of earthquake science, the idea that the whole field operates out of some constantly shifting middle ground between research and folklore, legend and fact. The same could be said of the New Madrid quakes, for these are temblors that have fallen, so to speak, between the cracks. Widely reported at the time of their occurrence—newspapers as far away as Washington, D.C., and Quebec City detailed local shaking, as the effects of the earthquakes reverberated across an area of almost a million square miles—they soon were relegated to the level of seismological footnotes, with no real place in the American disaster catalogue. It's tempting to explain this as a matter of chronology; many disasters of the nineteenth century (the burning of Boston, the San Francisco earthquakes of 1865 and 1868), have long since been forgotten, surfacing, if at all, as bits of historical trivia, like answers in a game of *Jeopardy!* Yet equally significant is the fact that, despite their size, the earthquakes were not destructive enough, in loss of life or property, to capture the imagination of their times. Although New Madrid sits only 180 miles south of St. Louis, it was pretty much uncharted territory in the early nineteenth century, inhabited by trappers and hunters, with no railroads, no communications network, nothing, really, to destroy. Add to that the long-held belief that the Mississippi Valley was not seismically active, and why would anyone remember New Madrid at all? If, as Hough suggests, the accepted view was that these earthquakes were anomalous, a one-time cluster of disruptions along a fault system that hadn't broken throughout recorded human history, there would be no urgency to account for them. In the meantime, the lack of substantial documentation would only make the whole subject that much easier to ignore.

It is here that folklore starts to become increasingly important, although to understand how, we need to go back to the early 1970s, when New Madrid first resurfaced as a matter of seismological concern. Around that time, a St. Louis University geologist named Otto Nuttli began to collect both eyewitness accounts and contemporaneous newspaper reports in the hope that, hidden somewhere between the lines of subjective experience, he might uncover a passage to the heart of the 1811–12 quakes. Relying partly on the efforts of researchers like early twentieth-century geologist Myron Fuller, and partly on his own historical detective work, Nuttli gathered about forty descriptions, or "felt reports," ranging from brief statements (the February 14, 1812, edition of *The Raleigh [North Carolina] Star* notes little more than the "smart shock of an earthquake" in relating the temblor of February 7) to elaborately detailed narratives. Based on these descriptions, he essentially reconstructed the earthquakes, developing the first comprehensive shake map of the region, and using it to assign each affected area a value from the modified Mercalli intensity scale—a twelve-point numeric system developed in 1902 by the Italian priest Giuseppi Mercalli that uses the physical effects of the shaking as a standard for interpreting the force of a quake. A Mercalli I earthquake, for instance, is "not felt except by a very few under especially favorable circumstances," while a Mercalli XII means "Damage total. Waves seen on ground surfaces. Lines of sight and level distorted. Objects thrown upwards into the air." If you're looking for a bit of perspective, the 1906 San Francisco earthquake measured, at its most punishing, Mercalli XI ("Few, if any masonry structures left standing. Bridges destroyed. Broad fissures in the ground. Underground pipelines completely out of service. Earth slumps and land slips in soft ground. Rails bent greatly"), while Northridge, for the most part, came in at Mercalli VII and VIII, with a maximum reported intensity of IX.

Nuttli's work on New Madrid provides yet more proof that, as with all things geopoetic, earthquake science relies on intuition, coincidence, personality even, to provoke certain necessary leaps. In 1973, after all, the year he released his findings in the *Bulletin of the Seismological Society of America,* the buzz in seismologic circles was all about plate tectonics, which had transformed the territory just a few years before. In such a landscape, it could be argued, no one but a geologist from St. Louis, a city rattled by two of the three New Madrid tremors, would even consider studying midcontinent earthquakes—what would be the point? "The plate tectonics revolution," says Susan Hough, "explained why we get this concentration of earthquakes at plate boundaries. As a result, things that happened between plates kind of fell off the back burner. And that pushed New Madrid further and further out of sight." Adding to this dissonance was the subjective quality of Nuttli's research, especially the Richter scale readings—7.2 for the initial earthquake in the sequence, 7.1 for the second, and 7.4 for the third—he extrapolated from the felt reports and their Mercalli values. Nuttli's work, in fact, was so inherently conjectural that throughout the three decades since his report's appearance, the New Madrid magnitudes have been continually revised, both upwards and downwards, on the basis of the same essential set of sources; one study, done in 1996 by Arch Johnston of the University of Memphis, argues that they could have been as high as 8.1, 7.8, and 8.0. In her own research, Hough has come up with readings much closer to Nuttli's originals, suggesting a magnitude of 7.2–7.3 for the first earthquake, with the latter two coming in at 7.0 and 7.4–7.5.

The irony, Hough believes, is that all this subjectivity may ultimately add up to something that, if not *objective* in any quantifiable sense, is in any case *measurable*—that is, able to be read. For her, the New Madrid felt reports (both the original forty used by Nuttli, and an additional sixty or so discovered subsequently) are astonishingly accurate, if you know what you're looking for. "The reliability of the accounts," she explains, "is the interesting question. There are a small number of detailed accounts, by a self-selected group of people who were writing down what they had seen. There was a physician and an engineer, a naturalist who happened to be in the area at the time; they were literate people, scientifically inclined. And importantly, I think, they had no preconceptions. So you have people describing things without an agenda, things they don't understand." Take the issue of amplification, in which the shock waves of a particular earthquake are exaggerated, or *amplified,* when they pass through loose soil sediment, such as we might find in a lake bed or riverbank. "There's one account," Hough notes, "saying that in Cincinnati the shaking was strong enough to swing windows back and forth and rattle cupboards, but then it goes on to say that away from the river, people slept through the earthquake. This gave me the idea that you had amplification along the river, which might have led to some inflated magnitudes." To test the hypothesis, Hough went back and looked for other reports of the phenomenon, only to discover it was a fairly common theme. "It turns out," she says, "that there are six or eight accounts that shaking was stronger in the river valleys than it was in the hills, which is completely in line with expectations now, but people then couldn't explain it."

The reason such observations are significant is that "they establish their own credibility, allowing us to get a sense of how internally consistent things are." In other words, they are reports we can take at face value, documentary narratives. As for why this matters . . . at the most basic level, it allows seismologists a more three-dimensional view of these earthquakes, a way to understand what happened at New Madrid, and to assess the area's current risk. Thus, just as Charles Lyall's reports of native legend, when correlated with modern physical data, give us a better idea of earthquake frequency in the region, so, too, do Hough's updated magnitudes and shake maps present a fuller picture of what we might expect to see when the Reelfoot Fault, "which all evidence suggests to have been the causative structure for this event," breaks again. This is especially important in regard to densely populated cities like St. Louis, Cincinnati, Memphis, even Mobile, Alabama—all of which were small settlements when they were shaken by New Madrid—and how they prepare for the inevitable moment when the earthquakes come again. "If these are magnitude eights going off every few hundred years," Hough says, "that starts to be a California level hazard. If you make them sevens, or low sevens, it makes a big difference in how we think. Take St. Louis, for example. If the shaking there had really been Mercalli VII or VIII, that would mean a big hazard for their masonry structures. But if it was intensity VI, the hazards are a lot more limited."

For all the practicality of Hough's research, there is another issue at work here, which has to do with how we look at earthquake lore. The New Madrid earthquakes have been the source of endless fables, from the not unreasonable supposition that they rang church bells in Boston, to the more improbable claim that they caused temporary waterfalls, or made the mighty Mississippi run backwards, south to north. Over the years, seismologists have written off the most extreme of these stories, just as they do many seismic legends, from earthquake smells like those Jack Lockhart notices to the cloud formations tracked by Zhonghao Shou. Yet paradoxically, Hough has found that when it comes to New Madrid, the more plausible the story, the less likely it is to be true. Although there are reliable reports of church bells ringing in places like Milledgeville, Georgia, there is no evidence that this happened in Boston; "The Boston papers," says Hough, "mention the earthquakes being felt in other areas, but never say they were felt in Boston, so the inference a historian would make is that they weren't felt there at all." In the meantime, anomalies like waterfalls or backwards-running water, while seemingly impossible, may actually have some basis in scientific fact. Such events, Hough points out, could have resulted from "uplift on the fault," a temporary raising of certain segments of the riverbed. Again, what's key are the accounts themselves, their apparent credibility, the way reading with a historian's eye for detail can indicate whether someone is telling the truth. Even John Clarke Edwards, who on February 19, 1812, published a letter in *The Pennsylvania Gazette* falsely claiming to have seen a new volcano, a "western Aetna" (the *Gazette* later printed a retraction), can't be entirely dismissed as a witness because of a previous letter in which some of his observations can be verified. "There are a lot of eyewitness accounts," Hough says, "that for a long time, the seismological community said, 'People are crazy, that can't happen.' But the more you look at it, the more the accounts have stood up. Now, I think, scientists have realized, 'Wait a minute, these people weren't crazy. These are valid observations.' "

Once we enter the realm of rivers reversing, of spontaneous waterfalls and volcanoes erupting, we're back in the psychic territory occupied by someone like Cloud Man, who seeks to walk a line between folklore and science, to take legend and alchemize it into truth. It's not the kind of landscape where you'd expect most seismologists to be comfortable, but for Hough, there's exhilaration in uncertainty. "One thing an earthquake like New Madrid has to offer us," she says, leaning across her desk, brown eyes glinting, "is the idea that anecdote can inform science, and some observations we see in anecdotal accounts we still don't understand." As an example, she cites precursory phenomena like earthquake lights or changes in groundwater levels before a temblor, which have appeared for centuries in eyewitness statements, but still have yet to be explained. "A lot of these reports," she tells me, "suggest you're either disrupting fluids or you're releasing gasses deep in the crust. Something happens—maybe hot gas comes to the surface—but it's not clear. In any case, something like that could explain a lot of anecdotal observations. And it inclines me to be open-minded about Cloud Man." Of course, being open-minded doesn't mean Hough believes Cloud Man's prediction system, necessarily, nor even in prediction as a general principle. Still, if her work on New Madrid has taught her anything, it's the need to wait and see. "He showed me some clouds," she says, "and they're not like clouds I've seen. He claims they're not normal clouds, although I don't know enough about clouds to tell. But what he's proposing is not scientifically outlandish, that the ground might release gas or vapor in the vicinity of a fault. He showed me his pictures, I gave him some feedback, and he's willing to go off in a sensible way. I'm not sure how good his scientific background is, but he's willing to do the work."

Listening to Hough, I feel myself start to drift again amid the endless push and pull of earthquakes, suspended between what we wish for and what we can say for sure. This is not at all what I expected; I came here to talk about science, after all. Yet what Hough most means to tell me is that, when it comes to earthquakes, science remains something of a question mark, a matter of reading between the lines. "Look," she says. "People debate whether or not earthquakes will ever be predicted. There's a school that says no, that it's a random, chaotic process, and you'll never be able to predict them, but I don't think we understand the system well enough to rule it out. If prediction ever happens, though, it'll be because someone realizes something we don't yet understand. It's going to come out of left field. So the people who are out in left field, like Cloud Man, maybe they're the ones who are going to discover something someday."

The notion that some major breakthrough in seismology might come out of left field is as compelling as it is unexpected, but it makes a lot of sense. When you get down to it, after all, the whole of seismology is a matter of left-field thinking, a whisper of a breath of an intuition, a momentary half-instant of thought. Seismology is a baby science, a newborn discipline relying on a mere hair's breadth of data, a field in which, at the most literal level, almost nothing is yet known. In seismological terms, the New Madrid earthquakes are important not just because they force us to make room for anecdote at the scientific table, but because they also mark, at a distance of 192 years, the absolute zero point of United States seismic history, the instant that our record keeping begins. There are tortoises and trees still alive that are older than this, cathedrals begun beforehand that remain unfinished, heritages still unraveled, dreams deferred. One hundred and ninety-two years is just a nanosecond, a blip in the consciousness of the universe, but when it comes to American earthquakes, it's the only time we have. Prior to that, we shift into the strata of earthquake prehistory, a shadowy epoch stretching back forever, past the Revolutionary Era and the Age of Exploration, far beyond Columbus and Cortez.

For most of us, the word *prehistoric* suggests something very specific, very human—that period of time before the onset of recorded history, before the Mesopotamians first scratched their ledgers into sheets of clay. With geology and seismology, however, prehistory takes on a different context or, more accurately, a pair of contexts that are by turns fundamentally related and inherently opposed. Geologic prehistory includes all the periods, ages, and eras of our lithic legacy, extending from the Cenozoic, which began 65 million years ago, down through the Mesozoic and Paleozoic, until, about 544 million years before the present, it falls off into the uncharted depths of the Precambrian, the nearly 3.5 billion years of earth chronology that predate the geologic time scale, and from which there are no records, no residue, nothing but the darkness of eternity. Seismologic prehistory, meanwhile, begins the day before the first earthquake was recorded, and it is different depending on the culture from which you come. The earliest known earthquake report dates from 1177 BC in China, although the Chinese earthquake catalogue is hardly continuous; even as seismically conscious a society as Japan has uninterrupted records going back just four hundred years. In the United States, *prehistoric* can refer to an event as recent as the massive quake—modern estimates put it at 9.0—that struck the Pacific Northwest on January 26, 1700, a catastrophe authenticated in the last few decades, after years of research led scientists to connect the dots between local Native American folklore, contemporary paleoseismologic evidence, and Japanese documents describing a large tsunami that made shore a few days later in Honshu.

What all this means is that, when it comes to earthquakes, the very concept of a scientific perspective is a fragmentary thing at best. It was only in 1883, for instance, that Grove Karl Gilbert, a charter member of the USGS and later its chief geologist, first posited in "A Theory of the Earthquakes of the Great Basin, with a Practical Application" that temblors might be caused by stress and slippage along cracks in the earth's crust, or what we now call faults. Gilbert's theory was inspired by an 1872 earthquake in Central California's Owens Valley, on the eastern front of the Sierra Nevada, which pretty much erased the town of Lone Pine, killing twenty-seven residents and destroying fifty-two of fifty-nine homes. In the aftermath of that event, writes T. A. Heppenheimer in his book *The Coming Quake*, "the fissures and faults in the earth were quite dramatic. In one place an area at least two hundred feet wide sank twenty to thirty feet, leaving vertical walls. Scarps or cliffs up to twenty-three feet tall formed along the eastern base of the mountains. Fences crossing the fault were broken, showing horizontal offsets of up to twenty feet." For Gilbert, operating at what was, in essence, the dawn of seismology, the Lone Pine earthquake seemed an inevitable side effect of the upthrust of the mountains' rising, a process he saw as triggering friction, and subsequently movement, on the fault. As he explains, "The upthrust produces a local strain in the crust, involving a certain amount of compression and distortion. . . . Suddenly, and almost instantaneously, there is an amount of motion sufficient to relieve the strain, and this is followed by a long period of quiet, during which the strain is gradually reimposed."

Gilbert's ideas on upthrust effectively illustrate the elusive mix of insight and observation upon which seismologists have relied for the past 120 years. Indeed, his analysis may be most startling because of its precision; leaving upthrust aside (earthquakes do help build mountains, but that's not what causes them), he's done a fine job of intuiting the basic dynamics of seismicity. In much the same way as Otto Nuttli or Susan Hough looking at New Madrid, however, Gilbert was limited by the lack of instrumentation to measure strain or shaking, or offer any independent explanation for what he'd seen. Although primitive seismometers had been employed in China since the Han Dynasty, the modern seismograph wasn't invented until 1885, when John Milne, an Englishman living in Japan, used a pendulum to amplify the at times virtually undetectable movements of the earth. By the turn of the century, Milne had installed a number of seismographs throughout Japan, creating the world's first seismographic network, although as late as 1904, when the city of San Francisco inquired about buying an updated version from the Japanese seismologist Fusakichi Omori, the only seismographs in California were still experimental models, not particularly reliable, installed at the University of California, Berkeley, and the Lick Observatory, near San Jose. Omori's seismograph, which San Francisco ultimately took a pass on (the asking price of two thousand dollars was deemed too high), pushed Milne's design a few steps further; among other things, it featured a rotating drum covered with photographic paper, on which the movements of an earthquake could be recorded with lines of light.

It's tempting to frame San Francisco's failure to acquire a seismograph as one of those great moments of historical irony: if only the equipment had been there, we tell ourselves, the city might have been able to prepare for the cataclysm of 1906. But the truth is that a seismometer would have had no effect on the 1906 earthquake, except, perhaps, to add a little more information to a documentary record that is already detailed and profound. Seismograph or no seismograph, the 1906 quake changed everything, not least the way we think about seismicity. Even as San Francisco's civic leaders sought to reframe the disaster for their own economic and political purposes, geologists were busy processing a vast array of new information that offered direct evidence of how earthquakes work. Within three days of the catastrophe, California governor George C. Pardee appointed the California Earthquake Investigation Committee, a panel chaired by Andrew C. Lawson of the University of California, and including not only Grove Karl Gilbert but Stanford University's John Casper Branner and Johns Hopkins's Harry Fielding Reid. Of these geologists, Lawson and Branner had, along with Gilbert, long been influential for their early work on faulting; Lawson is generally credited with having "discovered" the San Andreas, although, in actuality, the "infamous fracture" could not be discovered, since, as Caltech seismologist Kerry Sieh and Simon LeVay note in their book *The Earth in Turmoil,* it "lies in plain view—a dramatic gash in the landscape that runs more than half the length of the state." Reid, for his part, would introduce the elastic rebound theory as part of his work on the commission, an idea that Sieh and LeVay call "the basis of all subsequent thinking about earthquakes." To develop the theory, he compared nineteenth-century Bay Area land surveys with others made after the 1906 quake. What he found was evidence that, as Sieh and LeVay explain, "landmarks far to the west of the San Andreas fault (such as the lighthouse on the Farallon Islands, 25 miles offshore from the Golden Gate) were creeping gradually northward with respect to landmarks far to the east, such as the summit of Mount Diablo in Contra Costa County, 35 miles inland from San Francisco." As this movement continued, Reid concluded, strain, or "elastic deformation," would build up until it reached the breaking point, at which time an earthquake would occur. The terrain would then "rebound" to a "new, less strained position," and the cycle would begin again.

Looking backwards from the present, it doesn't seem like much of a leap from Gilbert's thoughts on faulting to Reid's elastic rebound theory. Certainly, the ideas are related, and only a quarter century or so separates them. In the brief history of seismology, however, the passage from one to the other is the difference between ancient history and the modern age. Gilbert's report on Lone Pine, after all, was almost entirely conjectural; although he was not present for the earthquake, he observed deformation in the region, and then extrapolated a rudimentary sense of what such patterns might mean. Reid, on the other hand, based his ideas on precisely recorded data, ranging from the survey information to the detailed mapping of the San Andreas that represents the other lasting legacy of the California Earthquake Investigation Committee. His resulting theory, while deceptively simple, was so far ahead of its time that its true impact would become apparent only more than fifty years later, with the advent of plate tectonics. In the meantime, seismology would undergo a series of slow but steady refinements, including the development of seismographic networks throughout both Northern and Southern California, increasingly comprehensive fault maps, the introduction of the Richter scale, and a more detailed recognition of the way faults broke.

Of course, the more attention geologists paid to California, the more the state became identified as earthquake country, precisely as the boosters feared. If, prior to the twentieth century, seismicity was a national phenomenon—besides New Madrid, a 7.0 decimated Charleston, South Carolina, on August 31, 1886, killing sixty people and collapsing more than a hundred unreinforced brick buildings—in the wake of San Francisco, there was literally a rewriting of American lithic history, in which catastrophic temblors *did not happen anywhere else*. Since 1906, more than a dozen California earthquakes have measured 6.0 or stronger, and that's just the high end of the spectrum; according to the Southern California Seismographic Network, something like two hundred thousand tremors altogether were reported between the years 1984 and 1994 in Southern California alone. There have also been, though, plenty of non-California earthquakes, like the 6.8 that shocked Seattle in February 2001. The largest earthquake ever to strike the United States, in fact, was epicentered not in California at all, but rather 120 miles east of Anchorage, Alaska, where a 9.2 shook for the unimaginable duration of three minutes on March 27, 1964. To look at pictures of that temblor is to see images so extreme they appear to have been staged for a disaster movie: houses fallen into enormous sinkholes, collapsed streets and sidewalks, buildings shaken off their moorings like so many scattered toys. This was a subduction zone earthquake, caused by one tectonic plate subducting, or pushing under, a second—a complex, volatile mechanism that produces intensely violent events. Once such a process was identified, the Alaska earthquake would significantly contribute to the acceptance of plate tectonics as a defining geologic principle, and, as such, help provide a larger context by which Reid's elastic rebound theory might finally be more fully understood.

If you want to wrap your mind around what Susan Hough calls "the plate tectonics revolution," perhaps the best approach is to label it the ultimate example of left-field thinking, a massive leap of intuition that paid off. Or maybe it would be more accurate to see it as its own variety of earthquake, an earthquake of the intellectual kind. Certainly, the idea was long enough in developing, the result of almost seismically slow shifts in the structure of geology, beginning with the sixteenth-century Flemish geographer Abraham Ortelius, whose 1596 *Thesaurus Geographicus*, John McPhee tells us in *Annals of the Former World*, "postulated that the American continents were 'torn away from Europe and Africa' by earthquakes and other catastrophic events." To Ortelius, the best evidence of such a process could be found in the shapes of different continental coastlines, which, in many cases, appeared to reflect each other like pieces of an enormous jigsaw puzzle spread across the surface of the earth. Look at any map of the world, and you can see exactly what Ortelius was referring to, the way that, say, the island of Madagascar looks like it would fit entirely into the gentle arcing curve of Tanzania and Kenya, while North and South America might close up like an iris with the jutting edge of western Africa, New York City pressed flat against Morocco, the Ivory Coast and Ghana mirroring the distant northern shoreline of Brazil.

For all Ortelius's foresight, it wasn't until just after Reid proposed his elastic rebound theory that the notion of continental drift, as the phenomenon came to be identified, was formally hypothesized by Alfred Wegener, a German meteorologist. Over the years, Wegener was roundly ridiculed, since despite the obvious physical evidence, geologists were hard-pressed to discover a mechanism that could explain such mass land movement across the globe. Then, in the early 1960s, a series of papers by researchers in a number of related disciplines began to develop what emerged as a unified set of theories, which, together, added up to a revolution in geologic thought. As McPhee describes it, "The story is that everything is moving, that the outlines of continents by and large have nothing to do with these motions, that 'continental drift' is actually a misnomer, that only the world picture according to Marco Polo makes sense in the old-time browns and greens and Rand McNally blues." This was plate tectonics, and instead of framing the earth as a fixed and ancient absolute, it gave us a geology based on change, on motion, in which twenty or so constantly shifting plates were responsible for *everything,* creating mountains, oceans, continents, as they moved throughout the limitless depths of time. Plate tectonics accounted for earthquakes—the faults that Gilbert and others had observed were plate boundaries, which meant that nearly all seismic activity could be interpreted as the result of two or more plates moving against one another—and what's more, it also explained the phenomenon of all those neatly matching continents, which at one time, or so the theory argued, had been part of a vast supercontinent called Pangaea. Seen in those terms, the earth, and the movement of the continents upon it, existed in a state of constant, organic evolution, much like a microcosm of the constantly expanding universe, with the first rupture of Pangaea the global equivalent of the Big Bang.

Once you start to consider plate tectonics in terms of the Big Bang, you run straight up against the central dichotomy of seismological thinking, which is, at heart, a dichotomy of time. When it comes to Pangaea, after all, we're talking about something 250 million years in the geologic distance, while plate tectonics was still being debated as recently as the early 1970s, although after 1968, Lucy Jones notes, "the majority of people realized that it just explained so much." These two very different conceptions of chronology—the geologic and the human—represent a pair of opposing principles that all seismologists have to keep in balance, as if they were looking through both the near and far end of a telescope at the same time. "Geology goes through twenty orders of magnitude of time from creation of the earth to plate motion," Jones tells me one afternoon in her large corner office at the USGS in Pasadena, just across the corridor from Susan Hough. "The youngest ocean is only 180 million years old—one twentieth out of the age of the earth." She laughs at the idea's dislocation. "Only 180 million years old, right? And then there are earthquakes: an earthquake happens in a matter of seconds, and we're dealing with speeds where a tenth of a second makes a real big difference. We time earthquakes to the nearest hundredth of a second or thousandth of a second. So you get to cover this incredible range, and you have to feel comfortable going back and forth."

Jones's fascination with geological processes coincides almost exactly with the mainstreaming of plate tectonics; in 1975, as a senior at Brown University, she took an introductory survey course and was compelled by what she saw as the field's integrated nature, the way it merged aesthetic, sociological, and scientific perspectives to seek an underpinning to the world. "It was an exciting time," she says, "when there were lots of questions and lots of answers, and everything you did, you came up with something. There was just a joy to what everybody was doing. Everywhere you turned, you'd figure it out. And all the questions sort of fit together." Given such a climate, Jones recalls, anything seemed possible, even earthquake prediction, which had become a priority among the seismological community. In May 1975—or at virtually the same moment Jones was first beginning to scratch her own geological surface—Frank Press, chairman of the Department of Earth and Planetary Sciences at MIT and perhaps the most prominent seismologist of his era, published an influential article in *Scientific American* claiming that reliable short- and long-term seismic forecasts were not only possible, but could well be commonplace within ten years. "The forecasting of catastrophe," Press declared, "is an ancient and respected occupation. It is only in recent years, however, that earthquake prediction has parted company with soothsaying and astrology to become a scientifically rigorous pursuit." Interestingly, Press's comments were uncannily similar to those made by Grove Karl Gilbert in a piece called "Earthquake Forecasting" that had appeared in the January 22, 1909, issue of the journal *Science*. "There was a time," Gilbert wrote then, "when the earthquake was equally enveloped in mystery, and was forecast in the enigmatic phrases of the astrologer and oracle; and now that it too has passed from the shadow of the occult to the light of knowledge, the people of the civilized earth—the lay clients of the seismologists—would be glad to know whether the time has yet come for a scientific forecast of the impending tremor."

The parallels between Press and Gilbert—down to specific points of language—suggest a certain unintended irony, which has to do with how little change, in spite of everything, sixty-six years brings. Even today, Jones acknowledges, the best-informed seismologist is hard-pressed to explain how earthquakes function; it's one of the great unresolved geologic mysteries, a major reason why many scientists now doubt prediction will ever be a workable tool. A quarter of a century ago, however, seismologists were captivated not only by the seismic shift of plate tectonics, but also by what appeared to be the first significant, documented prediction, of a 7.3 that hit the city of Haicheng, China, on February 4, 1975. Nine and a half hours before the shaking started, an alert was issued by the provincial government, which based its warning on a number of different factors, including several days of foreshocks (smaller precursory earthquakes throughout the region), as well as "changes in ground water which were usually changes in the level or color of the well water . . . [and] the appearance of a low ground fog." Most notorious among these precursors were many widely observed instances of anomalous animal activity, including snakes emerging from hibernation to freeze to death outside in the brutal winter climate, and packs of rats running along electrical and telephone wires. "Throughout January 1975," notes Helmut Tributsch in *When the Snakes Awake*,

> reports of unusual animal behavior kept coming into the central bureaus. More than twenty species of animals, among them snakes, rats, chicken, dogs, cats, horses, deer, and tigers, were said to have been seized by fear. At the beginning of February, the number of these reports suddenly climbed steeply. Increasingly, it was the larger animals, such as cattle, horses, and pigs, that were panicking. Different animals expressed this differently. Geese flew into trees, dogs barked as if mad, pigs bit each other or dug beneath the fences of their sties, chickens refused to go into their coops, cattle tore their halters and ran away, and rats appeared and acted as if drunk. Even three well-trained police dogs acted beyond all recognition. They refused to obey their handlers, howled, and kept their noses close to the ground as if sniffing something out.

Like the entire issue of earthquake prediction, Haicheng casts the unsettled relationship between scientific inquiry and anecdote in sharp relief. On the one hand, there's no doubt something happened; a warning was given, and although the city was not officially evacuated, many people left their homes, 50 percent of which collapsed during the ensuing quake. Even the most conservative observers agree that tens of thousands of lives were saved. Yet the real question, Jones and others insist, is not whether the earthquake was forecast, but whether the Haicheng prediction might be in any way repeatable, or was simply a one-time fluke, a coincidence, from which no lasting lessons could be learned. In the wake of the disaster, a team of ten American seismologists, including Jones and Christopher Scholz of the Lamont-Doherty Observatory, traveled to China to address these concerns, interviewing survivors and examining the evidence firsthand. What they found was less an organized prediction effort than a loose network of nonscientific systems, based in homes and schoolyards, where people had been told to look out for what were, in essence, traditional signifiers of the sort that had figured throughout centuries of earthquake lore. "It was a peculiar time," Scholz remembers, "because this was the Cultural Revolution, and Mao wanted to keep down the elites, and among the elites were scientists. So he picked a pet project to be done by the broad masses of people. Every little school in China had its own earthquake prediction project. The typical setup was that they'd have these little farmyard type things in schools, where they'd have ducks and chickens, and they'd observe them. But it wasn't just animals—they had all kinds of things. They had nails driven in trees, and they'd observe water wells, all these kinds of observational science."

Of course, the moment you start dealing with "observational science," you enter a territory that makes seismologists suspicious, the lessons of New Madrid aside. For Jones, the anecdotal quality of so much precursory information rendered Haicheng problematic, since it was impossible to determine with any certainly what had been observed. "Clearly," she and Peter Molnar stated in a report on the prediction inquiry to the American Geophysical Union, "the data are very crude. Only a tiny fraction were obtained by technically trained people. . . . Moreover, a large majority of the observations were obtained after the earthquake when people . . . from the Liaoning Provincial Seismology Bureau, the Institute of Geology of the State Seismology Bureau (SSB), and other work units of the SSB asked local people about possible precursors. This might have introduced a bias into the data particularly for the period immediately preceding the earthquake." Ultimately, the only precursors the American researchers could validate were the foreshocks, more than five hundred of which had rocked the area in the months before February 4. "Basically," Jones says, "what they did was guess this sudden swarm of earthquakes were foreshocks, and they were right." Still, she cautions, even that is far from conclusive, since foreshocks must be read retroactively, after the mainshock has struck. If you try to go the other way, you have no choice but to traffic in conjecture, which has been a hallmark of Chinese prediction all along. "There was another earthquake swarm six weeks earlier," Jones continues, "in a similar area, and it was evacuated, but nothing happened. That's the key to Chinese prediction—they're willing to have a very high rate of false alarms." Casting further doubt on the prediction program was the failure by Chinese officials to forecast the tremor that leveled Tangshan, one hundred miles east of Beijing, on July 28, 1976. According to Scholz, "Our committee was in Beijing six weeks before the Tangshan earthquake, and we asked if they were predicting anything. They said no. Just by virtue of us being there and asking those questions, they couldn't claim they had." Measuring 7.6 on the Richter scale, the temblor killed more than a quarter of a million people, the highest recorded death toll ever from a single quake.

In retrospect, it seems a bit surprising that the inconsistency of Chinese prediction did not undermine official interest in seismic forecasting. At the time, though, Haicheng was part of a much larger puzzle, one place among many where prediction was being studied around the globe. As far back as the 1960s,

Frank Press notes in *Scientific American,* Russian researchers "startled the seismological world with a report that unusual variations in the velocity of seismic waves appeared just before earthquakes in the Garm region of Tadzhikistan. Subsequently the Russians announced that in earthquake-epicenter regions in Garm, Tashkent and Kamchatka they had detected changes both in electrical resistivity and in the content of the radioactive gas radon in the water of deep wells." These findings would later dovetail nicely with much of what came out of China, but even before that, American geologists began to set up experiments to confirm the data, with some promising results. Eventually, a group of Stanford and Columbia University seismologists including Yash Aggarwal and Christopher Scholz came up with a prediction model, the dilatancy-diffusion theory, which was developed in 1972 and 1973. The basic idea was that, in the period just before an earthquake, small cracks would appear in an area of developing strain. Such cracks, in turn, would alter the physical properties of the rock along the fault. Partly, as the Russians had noted, this involved the velocity of seismic waves, but equally significant was a shift in water flow, which swelled, or *dilated,* rocks in the affected zone. As the earthquake drew nearer, water would *diffuse* back into the cracks, "weakening the rock," Press explains, "to the point where small earthquakes increase in number and the main shock follows."

Initially, dilatancy-diffusion looked like the long-awaited Rosetta stone of earthquake prediction; it incorporated a wide range of well-known precursory phenomena, from foreshocks to electrical anomalies to changes in water level, and had the added imprimatur of scientific authority, as well. The theory's stock went up only when, on August 2, 1973, Yash Aggarwal noticed variations in the seismic velocity near Blue Mountain Lake, in upstate New York, and announced that a small earthquake (magnitude 2.5–3.0) would take place there the following day. The next evening, a 2.6 was measured in the region, which was immediately interpreted as a validation of Aggarwal's work. In such an atmosphere, even the events in China could be read as something of a great leap forward, all the inconsistencies little more than ruffled edges to be ironed out as the available information on prediction grew. This sense of possibility became more pronounced only with the discovery, in early 1976, of what came to be known as the Palmdale Bulge—an uplift of more than eight inches along the San Andreas Fault in Palmdale, a small Los Angeles County city about seventy miles north of downtown L.A. A couple of factors made the bulge seem particularly significant. First, it sat along a stretch of the fault that had last slipped during the Fort Tejon earthquake, more than a hundred years before, and had been, as Carl-Henry Geschwind points out in his history *California Earthquakes*, "ominously quiet and free of even small earthquakes since at least 1932." Equally important, its existence seemed to support a hypothesis made by Christopher Scholz three years earlier, in which he argued that if rocks dilated before an earthquake, the increase in their volume should cause the ground to bulge. Shortly after the bulge was identified, James Whitcomb, a seismologist at Caltech, turned the heat up further by forecasting that an earthquake measuring between 5.5 and 6.5 would strike greater Los Angeles in the next year. Bills were introduced in Congress to authorize additional research funding. It looked as though prediction had arrived.

And then . . . nothing happened. Not only did Whitcomb's earthquake fail to materialize, a suit was filed by homeowners who claimed the false prediction had damaged property values. "It got dropped," Lucy Jones says, her voice dripping disdain, "but it was classic L.A." The matter of the Palmdale Bulge, meanwhile, seemed to dissipate of its own accord. "By 1979," Geschwind writes, "most of the bulge had deflated without any release of strain in a great earthquake. In fact, a number of geophysicists came to argue that the bulge had never existed and that the appearance of an uplift near Palmdale had been due to measurement errors rather than real movement of the earth's surface." Dilatancy-diffusion fared no better. In 1978, Allan Lindh of the USGS's Menlo Park office found that much of the supporting data had been incorrectly interpreted, which raised considerable questions about the theory as a whole. Most troubling were certain unresolved contradictions between the stresses needed to produce dilatancy and the failure strength of a fault. "The most important discrepancy," suggests Tom Heaton, a former USGS geologist who now works at Caltech, "was the discrepancy of scale and size. If you take a piece of rock in a laboratory, and you take it up to the point where it fails, it stores tremendous energy. You can't be next to a piece of rock that's confined and breaks in a laboratory because it's extremely dangerous. If earthquakes were like that, we'd all be killed instantly, as soon as the fault broke." Faced with this, the scientific community backpedaled on forecasting, confounded one more time by the enigmatic nature of earthquakes, the way they never seem to do what they're supposed to, almost as if they have minds of their own. During the 1980s, the USGS took one last stab at prediction, setting up an extensive network of monitoring equipment in Parkfield, a town in Central California where moderate temblors had been observed at what appeared to be regular intervals. But although, between 1988 and 1993, the Survey dutifully declared a series of earthquake warnings, not a single one ever panned out.

The USGS experience at Parkfield raises a lot of issues about the, er, unpredictability of earthquake prediction, but once again, the most fundamental has to do with time. In human terms, it's not hard to see how Parkfield came to be considered something of a seismic alarm clock, since it seemed to go off with a consistency unique within the realm of earthquakes, a consistency that suggested subterranean connections, a larger lithic order that had chosen this place to assert itself. In 1857, 1881, 1901, 1922, 1934, and 1966, the town, which sits not far from the San Andreas, was shaken by magnitude six-plus earthquakes; the last two in the sequence were also preceded by 5.0 foreshocks seventeen minutes before the main quakes hit. "Six earthquakes," write Kerry Sieh and Simon LeVay in *The Earth in Turmoil*, "all pretty much alike, at an average interval of 22 years. And if you're willing to say that the 1934 event 'came early,' then you can make the case that there's some underlying clockworklike mechanism that sticks to the 22 year cycle quite closely."

The problem with Parkfield, however, is that earthquakes don't come early (or late, for that matter); that's an inferential overlay, a piece of observational science taken to extremes. "The Parkfield predictions," says Tom Heaton, "were based on what looked like a pattern. The idea was, we see the pattern of this repeating earthquake and we're going to predict something in the future. In the statistics business, that's called the Texas sharpshooter problem—if you look for patterns, you will see them. If you then do a statistics test, the statistics support you, because you made the rules after you started the game." A perfect case in point is the ten-year fudge to justify the 1934 earthquake. "They made an assumption," Christopher Scholz laughs. "The one earthquake that happened out of sequence, they threw it away. If you leave it in, your window of uncertainty is much wider." This uncertainty becomes even more pronounced if we continue to go backwards, beyond the narrow confines of our epoch, and into geologic time. Although 1857 may return us to the early years of U.S. earthquake history, it's far too recent, geologically, to reveal anything definitive about the seismic legacy of the region, let alone what the future has in store. Here, then, as much as anywhere, the patterns we confront are of our own creation, what, in this instance, we may call the mismeasurements of man.

Nevertheless, despite the complications that arose in Parkfield, it's not the concept of a pattern that is necessarily fallacious, just the available sample that was too small. What geologists needed was a way to stretch the boundaries of seismic history, to gather information from beyond the range of what we might commonly be said to know. Tellingly, even before the Parkfield experiment was set in motion, Kerry Sieh, among others, had started work on just such a process by excavating a few small sections of the San Andreas along the stretch where it traverses the Carrizo Plain. The Carrizo Plain is as close as Southern California has to a natural seismic laboratory, "a hot, dry uninviting piece of real estate tucked in between the Coast Ranges and the aptly named Temblor Range, about 120 miles northwest of Los Angeles," which, by virtue of its desolation, remains a relatively pristine environment, where the fault can be observed quite clearly, unfolding across the landscape like an attenuated seam. Sieh's idea was to look to the San Andreas itself for physical evidence of events predating the 1857 Fort Tejon earthquake, using a modified archaeological approach. Unlike Otto Nuttli and the researchers of New Madrid, he was less interested in anecdotal data—although he took note of it where it was available—than he was in issues of offset and radiocarbon dating, by which he hoped to reconstruct some of these prehistoric ruptures on the fault. In the fall of 1975, Sieh began his investigations at a site called Wallace Creek. But it wasn't until later in the year, when he turned his attention to the similarly named Pallett

Creek, thirty-five miles northeast of Los Angeles, that he stumbled across an unexpected bit of information that would ultimately change the way we think about seismicity by expanding our frame of reference to a depth of several thousand years.

For Sieh, the key discovery was a thirty-foot-deep gorge the creek had sliced across the San Andreas, revealing layers of peat and other sediment, in much the same way that a highway roadcut does. Earth scientists have long considered roadcuts to be something like the individual pages of a massive lithic history, a geologic narrative of revelatory force. "The roadcut," John McPhee writes, "is a portal, a fragment of a regional story, a proscenium arch that leads their imaginations into the earth and through the surrounding terrane." With a similar sort of intention, Sieh decided to study the deposits in the Pallett Creek gorge, to see if he might map disruptions on the "fault trace" over time. Once he began to work in earnest, what he found astonished him: an archaeological record of a dozen major earthquakes on the San Andreas, extending back from Fort Tejon in 1857 nearly to the time of Christ. If you look at photographs of Sieh's Pallett Creek excavations, you can see the traces of this history very clearly; there are well-defined breaks in the layering of sediment, where continuous veins of rock or other material are abruptly ruptured and uplifted, as if the pieces of a puzzle had been pulled apart. In some sense, it's like looking at the outlines of continents and then trying to figure out how they fit together, but at the same time, it's less abstract than that, more like an earthquake you can touch. Twenty-three years after Sieh first began digging at Pallett Creek, I stood in a trench cut through the very middle of the San Andreas a hundred or so miles to the southeast, near Redlands, and watched as another seismologist, Sally McGill, marked patterns of dislocation with small red flags.

The day was hot and still, as it always seems to be around earthquakes, and I couldn't help but wonder what would happen if one struck.

"Well," McGill told me when I asked her, "it depends on where you're standing, and where the rupture is. If it happens here"—she pointed to an imaginary line between us—"you'll probably end up twenty feet down that way. But if it happens there"—and now she pointed to a second imaginary line, this one running underneath my body—"then," she laughed, "you're out of luck."

I want to tell you that I was scared by this, that I scrambled up out of that trench like a wild animal, that I ran to my car and drove away from the fault without ever looking back. But the truth is that I felt much the same way standing in the center of the San Andreas as I do when faced with those images of Pallett Creek. In both cases, it is not fear that resonates, but wonder, the promise that, if we could simply see it from the inside, this world in all its mystery might cohere. This is a sensation I often have in California, as if I am never more than a breath away from the elemental, as if anything could happen at any time. Occasionally, staring at Sieh's pictures, I brush my fingers across the photographic paper, just as, that afternoon with McGill, I ran my hands along the edges of the fault itself. This thread of rock tore in 1680, I remember telling myself, and this in 1480, and this in 1100, barely a dozen years after the Norman Invasion, and a full century before the Magna Carta was signed. I felt the years collapse, felt time elide beneath my grasp. It was as close to the eternal as I've ever been.

Of course, the irony of Sieh's and McGill's research—not to mention the geologic subspecialty of paleoseismology it has engendered—is that even as it brings us face to face with the eternal, it highlights just how distant this eternity remains. On the one hand, Sieh's discovery of ancient earthquakes enabled him to construct a time line, a pattern of breaks along the San Andreas over a period of two thousand years. If this is not quite the territory of geologic time, it's a lot closer than, say, the span of years involved at Parkfield, and it increased our understanding of the mechanics of the fault. Based on the paleoseismological evidence, Sieh was able to conclude, with a fair degree of confidence, that the southern segment of the San Andreas breaks, on the average, every century to century and a half. "What we found," he explains, early one fall day in his office at Caltech, "was that at the three or four sites where there's a good record, we have a 105- to 160-year average interval. It varies from site to site, so near Wrightwood, for example, the interval seems to be about 105 years on average. That's how often it breaks. It's now been 145 since the last one. At Pallett Creek, where I started the work, 132 years is the average; it's been 145. Halfway up to San Francisco or so, it's an interval of 160, and 145 since the last one."

Yet if calculations like these imply a certain logic, an overriding structure to the ebb and flow of earthquakes, such an order, Sieh insists, is too vague to do more than suggest the broadest patterns, which only illustrates the way seismicity continues to confound us, to defy our efforts to put forth a framework that makes consistent sense. The problem with an average, after all, is precisely that: it's an average, and over the course of two millennia, the southern San Andreas has yielded some significant anomalies. For every interval that reflects the median—the 132 years between 1680 and 1812, or the 134 that stretch from 1346 to 1480—others have been far greater or smaller, reminding us of just how little about earthquakes we truly know. More than two centuries, for instance, elapsed between a massive temblor in 1100 and the 1346 event, which broke a shorter segment of the San Andreas; a similar span divides the quakes of 1480 and 1680, the latter of which ruptured the fault only south of San Bernardino, and didn't even register at Pallett Creek. Alternately, less than forty-five years separate Fort Tejon from the tremor that immediately preceded it, which leveled the church at the mission of San Juan Capistrano on December 8, 1812. "After 1812," Sieh laughs, "you wipe your brow and say, I'm glad that's over. You know, we just had five meters of slip, and we're probably good for the next 150 years. Then, 45 years later, another great earthquake happens, and you think, Wow, that was a hundred years early. What's going on? It reminds me of a quote by the historian Henry Adams, who once said, 'Order is the dream of man, and chaos is the rule of nature.' This gets to the heart of what we're dealing with in science, and also what we're dealing with in our lives, in our culture. The intervals are too big, too sporadic, for us to predict anything with certainty. But we have made inroads into understanding how the planet works."

In the end, this tension between form and chaos, between what these patterns may tell us and what they may not ever tell us, reflects the essential tension at the heart of the earth sciences, the tension between pragmatics and philosophy. If the pinpointing of prehistoric earthquakes opened up the territory of seismology, allowing geologists access to a broad new strata of information, it also led them inexorably back to the slender surface layer of the present, because the past remains so fluid, so amorphous, it's virtually impossible to get at, except in terms of probabilities. "I sometimes use an analogy comparing seismicity with weather," Lucy Jones explained once when I asked about it. "You have climate, which is what you expect in the long-term history of a place. And you have weather, which is what you're predicting for tomorrow. We can do earthquake climate. Give me a hundred thousand years, and I'll tell you what earthquakes are going to happen. That's not random. That's deterministic, and it's driven by plate tectonics. Which subset of them are going to happen in our lifetime is completely random. If we go back to the climate-weather model, the climate isn't random; the weather is. The random part is when you try to do it in a human time frame." Ask Sieh the same question, and he'll suggest a slightly different posture. "Before I started my work in paleoseismology," he says, "nobody knew how often the San Andreas broke. All we knew was that, historically, there were two big events, 1857 down here and 1906 up in the Bay Area. Were earthquakes expected every thousand years, ten thousand years, hundred years? Nobody really knew. So if you know what happened in the past and how frequently, you not only know the magnitude of what might happen in the future, you know whether it's likely or not."

When Sieh refers to the potential circumstances of some future earthquake, what he's really talking about is hazard management, which, throughout the 1990s, has increasingly become a focus of the field. On some level, this is entirely understandable—that, faced with the earth's intractability, many seismologists would turn away from larger questions, becoming less theoretical, less intuitive, more directed towards the here and now. "The idea that seismology isn't a valid scientific endeavor because we can't predict earthquakes is misguided," Susan Hough says, with no small measure of annoyance. "It's a bogus argument because there are plenty of testable hypotheses that involve assessing future hazards, future ground motion on active faults." Getting a grip on ground motion, in fact, is now a central goal of earthquake research, as seismologists begin to pay attention less to when the next event will strike, exactly, than to what will happen when it does. Since the late 1980s, the USGS, Caltech, and the California Division of Mines and Geology (recently renamed the California Geological Survey) have collaborated on a project to install state-of-the-art seismometers across Southern California, a network that has helped emergency responders react more quickly to disaster, since they can immediately determine what areas have been hardest hit. Originally known as TriNet, the program has been reinvented also—as the California Integrated Seismic Network, or CISN—and expanded throughout the state. For all its success, there's nothing particularly revolutionary about an idea like CISN; in many ways, it's a contemporary analogue to the seismographic network that John Milne established at the end of the nineteenth century in Japan. But if this makes for a stunning symbol of circularity, it also reiterates ever more deeply just how short a distance seismology has really come.

As for the future, there is talk about someday using the CISN as the basis of a limited early warning system, in which, once an earthquake starts, information would be relayed throughout the region, enabling local municipalities and utilities to shut down potentially sensitive or at-risk operations—nuclear reactors, elevators, subways, commuter trains. The Japanese are already experimenting with this type of program, as are authorities in Mexico City, who hope to mitigate another 1985-style earthquake, which began with a fault slip more than two hundred miles away. "Theoretically," says Lucy Jones, "it's a very straightforward idea because when the fault surface generates the earthquake, you start at one point, and the rupture propagates from there. That process can take up to two minutes, so you have quite a bit of time from when the earthquake starts. Southern California is actually very well suited to make use of something like this because our major fault is outside the city. So we have travel time." Still, Jones cautions, even if such a system were available, we'd be better off concentrating on our own homes, our own safety, all the mundane daily responsibilities of living in an earthquake zone. "What we need," she argues, "is a way to live with earthquakes. We need to build buildings that won't fall down. We need to be able to deploy the necessary resources, even after an earthquake takes place. Whether or not we can predict them, earthquakes are going to come. This is something you live with by preparing, and we need to be prepared."

Listening to Jones, I find myself struck by the common sense in what she's saying, her calm, considered way of taking stock. Certainly, preparation is the most important task we can undertake in regard to earthquakes, more important than science, even, or any of the stories we tell ourselves. At the same time, preparation on its own is not enough. How could it be, when even now seismology continues to reconstitute itself, reframing the ideas at its core? When I first met seismologists like Jones and Tom Heaton in 1998, it was commonly believed that there was no real difference between how large and small earthquakes started, an idea that seemed to invalidate the notion that you could ever predict an earthquake's size. Back then, seismologists believed earthquakes grew (or didn't) as they propagated, which meant the only real determinant of magnitude was the length of a rupture on the fault. Plenty of geologists still think like that, but in the last few years, an associate professor of geophysics at Stanford named Greg Beroza has begun to produce research that suggests this random onset model might not be so accurate, after all. "Greg's definitely come up with something," Jones says now. "It's an open question. I'm less certain that earthquakes begin randomly than I was a few years ago." In the brief history of seismology, this can be read as yet another symbol of all the unresolved (and, perhaps, unresolvable) questions that reside at the center of every earthquake.

For me, all this points up the final, irreconcilable problem with the pragmatic; it's too small, too narrow, too out of balance, too unwilling, ultimately, to be wrong. Hazard assessment may be essential, but there's no risk in it—by its very nature, it is risk averse. Geology, however, is a risky business, and if you look back, all the great discoveries have been outgrowths of someone taking chances, from Grove Karl Gilbert, with his exquisite stretch of the imagination about faulting, to Alfred Wegener and his theory of continental drift. Even Kerry Sieh's decision to excavate the San Andreas was a plunge into what was, at the time, almost entirely uncharted territory, in which he stumbled upon his most important realizations by mistake. In every one of these instances, we are in the realm of left-field thinking, of leaps of faith, of inspiration, of *geopoetry*. We are in a landscape, in other words, where eternity presses up against the present, where, for a fleeting moment anyway, we can hold them both simultaneously in our minds. It is this that has always attracted me to earthquakes, this sense of bottomless reality, this possibility. Even to think about it feels like standing in a trench dug across the fault line, fingering the residue of history.

Several years ago, not long after I went to see Sally McGill out at the San Andreas, I came in contact with a predictor named Jim Berkland, who lives in Sonoma County. Berkland is best known not only for forecasting the Loma Prieta earthquake, but for getting it on record; on October 13, 1989, four days before the temblor struck, he was profiled in the *Gilroy Dispatch* for having predicted a "World Series Quake." A maverick seismologist—he worked for the USGS from 1958 until 1964, then spent five years at the U.S. Bureau of Reclamation before he was hired as a Santa Clara County geologist in the early 1970s—he based his predictions on a combination of lunar tides and anomalous animal behavior, and the first time I ever talked to him, in late 1998, he told me "with 80 percent confidence" that a 3.5–5.5 magnitude earthquake would strike within 140 miles of San Jose between December 1 and December 8. On December 4, it happened: a

4.1 in Berkeley, the strongest quake to hit the Bay Area in four months. When I called back, he said we were in an active window, and I should expect a 4.0–6.0 in Los Angeles before December 10. By way of evidence, he mentioned that seventy-two dogs had recently gone missing in Southern California, as compared with fifty-eight in the days before the Northridge quake. As we spoke, I felt a tingle of excitement, like I'd been promised something rare and beautiful, something I didn't know exactly how to see. It wasn't that I was hoping for an earthquake, but there *was* a part of me that wanted to believe.

Then, two nights before the close of Berkland's window, I was driving home along the broad rolling curves of Sunset Boulevard, heading east through Beverly Hills on my way towards West Hollywood. On the radio, R.E.M.'s "Man on the Moon" played like a sound track, and as I stared at the office towers silhouetted against the edge of the Strip, I started to think about the fault that ran beneath this pavement, wondering what would happen if it slipped. Cresting the small hill at Doheny, I caught sight of the moon, hanging fat as a cocktail onion, low and close in the sky. It was so big that it almost filled my windshield, so big I could nearly feel its pull. This, I imagined, was what Berkland had been trying to explain, this interaction of moon and earth, and for a moment, I could almost see his words in action, see the moon on the closest part of its orbit, exerting its tidal pull. As I considered this, "Man on the Moon" faded into a series of tight, martial drumrolls, and Michael Stipe started singing, "That's great, it starts with an earthquake"—the first line of "It's the End of the World As We Know It (And I Feel Fine)." All of a sudden, I felt like I'd been given a set of signs, like a trapdoor had opened to expose the real California, the wild and elemental territory of our nightmares and our dreams. I looked around: life went on as normal. Club kids hung out in front of the Rainbow and the Roxy, while traffic moved past on Sunset at a crawl. In my head, though, it was as if reality itself had started to slip, as if somewhere out on the boulevard, I'd been put in touch with some kind of strange, intuitive logic, and it was telling me tonight's the night.

I finished the drive in a weird state of heightened awareness, registering every bump in the road, every gust of wind. Even after I got home, the sensation lingered, and I moved from room to room making sure cabinets were closed, moving anything that looked like it could fall on my sleeping children's heads. On some level, I knew, this was ridiculous, a classic case of the power of suggestion overwhelming the power of mind. Yet the edge I was feeling grew only more acute when I checked my e-mail and found an update from Charlotte King, who cited "Heart pain . . . on and off the last few hours," a precursor (or so she said) to activity in Yucca Valley, Landers, or Big Bear. "Whatever is happening that I am picking up," King wrote, "will be happening in less than 12–72 hours . . . more likely 12–24 hours. We are looking at a moderate size event, 4.0–4.6+." The message seemed to confirm Berkland's prediction, which, in turn, only solidified my own aura of belief. It didn't matter whether all this was a matter of magical thinking; it didn't matter if it was true or not, just that it *might* be.

Berkland's earthquake never happened, and that, seismologists will tell you, is all you need to know. On a purely pragmatic level, they're right, but what I keep coming back to is my sense of wonder, which continues to resonate within me even now. It is this I most remember about that December night, the way that, sitting in my living room, pondering whatever might be coming, I found myself once more face to face with infinity. I closed my eyes and visualized the breaking fault, whole strata of the past, of centuries, compressed and sliding together, waves radiating out like echoes of prehistory. I imagined how these waves would sweep into the present, how they'd bring the wood beams of my house alive. I felt the empty, stomach-dropping awe that reverberates each time I contemplate such movements of the earth—the awe of time, of inexorable forces, an awe that touches me more deeply than any practicality. Sometimes, I knew, this awe is all that will sustain us when the milliseconds yield to millennia, and still the world remains a mystery. Then I settled in, and waited for the shaking to start.

EARTHQUAKE COUNTRY

To drive north out of Los Angeles is, in essence, to see the seismic history of Southern California unfold. It's a physical—no, a *meta*physical—experience, as if I've entered a three-dimensional flip-book, in which I slip backwards from the present step by step into the past. From my house, I head south on La Cienega Boulevard half a mile or so until I reach the on-ramp for the 10. Here, in 1994, the freeway overpass buckled during the Northridge earthquake, leaving an entire section of roadway bent and bowed above the surface streets, so low in places that, if you happened to walk underneath, you could reach up and touch it with your hand. For me, this site has long held a kind of metaphoric urgency, like a message meant to tell me just how disastrous my own situation might become. I drive beneath this overpass every day, sometimes two or three times a day; it's a divide that may, at some point, separate me from my children, since both of them go to school on the other side of that line. Each time I have to wait beneath the freeway for the light to change, each time there is a traffic jam, or an accident, I think (if that's even the word for it) how much I do not want to die down here. During both of the last two big urban California earthquakes—Northridge and Loma Prieta—people's lives ended in just this fashion, with the shearing of concrete and steel support beams and the shattering of asphalt, with the world collapsing not from underneath them, but from up above.

From La Cienega, I take the 10 west to the 405, crossing the flats of West L.A. Then, I angle north, over the Sepulveda Pass and up into the San Fernando Valley, which, along with Santa Monica (where the loose sediment soil amplified the temblor's shock waves, in much the same way as happened along the Mississippi River during the New Madrid sequence almost two hundred years before), suffered the most significant damage in the Northridge quake. The 405 takes me through Sherman Oaks, where houses slid off the backsides of the Hollywood Hills, leaving cars and bodies buried in debris; later, it passes about three miles east of the intersection of Reseda Boulevard and Saticoy Street, where, according to a computer-generated shake map prepared by Kerry Sieh and Egill Hauksson at Caltech, the earthquake had its epicenter on a previously unmarked thrust fault. By the time I merge onto the 5 freeway, I'm nearly in Sylmar, site of the 6.6 San Fernando tremor, which killed fifty-eight people on February 9, 1971, destroying the San Fernando Veterans Administration Hospital, among other structures, and seriously damaging the half-century-old Van Norman Dam. Less than ten miles up the road, I pass through the Clarence Wayne Dean Memorial Interchange, named for the motorcycle policeman who died in this very location, within sight of the northernmost boundary of the City of Los Angeles, where the Antelope Valley Freeway sweeps into the 5.

The Clarence Wayne Dean Memorial Interchange makes a strange border crossing for a variety of reasons, not least the absurdity of naming a freeway exchange for a human being. It's a classically Southern Californian gesture, one that would make little sense elsewhere, yet somehow, it compels me just the same. It's not only that a man died here, but also the nature of how we live in the Southland, how we interact with our environment, which makes freeways among the most important monuments we have. If earthquakes are one symbol of the California experience, then cars, speed, tarmac all make up another, the key to every cliché you've ever heard about reinvention, about this state being a place where people come to let go of the past. One of the biggest challenges facing contemporary geologists, actually, is the degree to which California is continually being developed; all those malls and streets and housing complexes cover up the seismic record, literally erasing any vestige of what came before. In that sense, it's a striking juxtaposition, this intersection of road and earthquake, surface and depth, endless present and endless history, and it's made all the more so by the fact that every time I drive by this spot, I can't help but think about how it looked in the first photographs after Northridge, connectors shattered, rebar and rubble strewn across the asphalt, whole sections of roadway cast aside like loosely scattered cards. This double vision is what haunts me, what follows me some days even more closely than my shadow, this sense that I am living on a tightrope, traveling a line between the intractable past and the unknowable future, with only my memories and intuitions to serve me as a guide.

I continue north, through Valencia and Castaic, past the swooping roller coaster curves of Magic Mountain and the twisted wreckage of the San Francisquito Dam. Up ahead, the vista of the road begins to narrow as it rises through the hills of the Angeles National Forest, dun-colored and dry beneath the mid-October Friday morning sun. The further I climb, the more the radio fades into a whining static, which is just as well, since today it's all anthrax and Afghanistan. Listening to it is like going through another kind of earthquake, a psychic one that never ends. Just five days ago, on the Sunday that the U.S. air campaign started, I was home with the kids when Rae called to let me know the news. Immediately, I went to see how everyone was doing: Noah was in his room, playing, but I found my three-year-old daughter, Sophie, sitting in the living room watching a Winnie-the-Pooh video, talking to the characters in a whisper reedy as the breeze. Briefly, I felt my heart seize at the sheer simplicity of the moment, the unexpected way the mundane sometimes yields to the miraculous; beyond these walls, the world might be falling into chaos, but here, everything continued as it always had. No sooner had I slipped into the room, however, than I felt a quick jolt, a telltale rumbling, a thrum of wood beam slapping against wood beam. Great, I thought, the perfect moment for an earthquake. Then, I called to Noah as I gathered Sophie up against me, and in a low voice, almost unconscious, muttered, "It's like the end of the world."

Of course, if geology has anything to teach us, it's that the end of the world is never really the end of the world, at least not in the global sense. That Sunday earthquake was little more than a hiccup, barely a 2.8, the very bottom range of what is felt. Even had it been significant, there is always another iteration, another continent, another plate shift, always another way by which the planet proves itself to be elastic, able to spring back. Within the narrow span of human history, the end of the world remains wholly relative, as well. In 1857, the Fort Tejon earthquake represented as definitive a point of closure as many people could imagine; "When I was a child," a Native American woman remembered nearly sixty years afterwards, in a report cited by Philip L. Fradkin, "there was an earthquake; the earth shook and the ground cracked in Cholame. We were frightened and thought that the end of the world had come." This morning, though, as I crest the long, slow upgrade of the 5 near Gorman, I see signs for the Fort Tejon State Historical Monument, where everything still stands, much as it did during the time of the earthquake, here at the top of Tejon Pass. This is the spot where the fault and the freeway cross each other, just before you slide down through the Grapevine and out along the endless horizontal platter of the Great Central Valley, which stretches like some vast stage set on the other side of this ridge. For a moment, I consider pulling off the road to see the monument, but I've got a long drive ahead of me, and anyway, the Tejon Pass is not a very hospitable landscape, exposed to high winds that buffet my car even on a day as clear as this one, and loud with the groaning gears and engines of eighteen-wheelers as their drivers strain to push them over the hill.

In many ways, it's only fitting that all these earthquakes should add up to something, even if it's only something I've invented: a line of passage leading out of Southern California, like an arrow pointing its long and rocky finger north. This is what we mean when we say we look for patterns, and today, I'm after patterns on a grand scale, heading up to the Bay Area, where, I hope, the chasm I've been straddling between practicality and prediction, the mundane and the miraculous, might solidify into a shape that I can see. Among the reasons for my trip is to visit Jim Berkland, who, as he has for years now, continues to confound. The last time I spoke with him, a little more than a month ago, he mentioned that the Mayan god of earthquakes was named Olin, which can be taken as a sign or a coincidence, depending on your point of view. Then, he went on to note that we were in the middle of another seismic window, that he was calling for a 3.5–6.0 in Southern California within the next five days. This time, I didn't give much credence to his forecast—until, that is, forty-eight hours later, when, as I tried to teach Noah how to twist his fingers around a guitar neck in the shape of a G chord, we were shaken by a pair of earthquakes: a 3.1, followed within a matter of minutes by a second, smaller shock. Too slight, yes, by a fair margin, and one too many temblors to fit Berkland's model. But the timing was nothing if not eerie, a reminder of the leaps of faith, of intuition, which even now I can't resolve.

As much as anything, Berkland is an emblem of this lack of resolution, a living symbol of the contradictions that, for me at least, earthquakes represent. Towards the end of our most recent conversation, he boasted that, out of eleven major temblors thus far this year, he had successfully predicted eight, but when I pressed him for the details, he veered off, as is his tendency, into an entirely unrelated speculative landscape, recounting a trip he had taken to Machu Picchu, where he'd meditated in the King's Chamber, and, in the process, discovered (or so he told me) the secret of life. Statements like this have always made me wary about Berkland. For all that his predictions interest me, I can't help hearing in his voice the conspiracist's fine, hard edge of obsession, as if he were a latter-day William Money preaching imprecations in the wind. Perhaps, like Charlotte King, he can't control himself. Or maybe such single-mindedness comes with the territory, and after years of being dismissed, or worse, debunked, he's developed his own way of speaking, a fast and jagged monologue designed to circulate as many ideas as possible before someone shouts him down. Either way, I feel my body clench when Berkland starts to talk like this, as if he's taking me somewhere I'm not prepared to go. For all that they put us in touch with the eternal, I don't believe earthquakes offer secret messages from the universe, at least not in the abstract manner Berkland means it; if there's a philosophy to seismicity, it is a physical philosophy, one whose tenets play out in the ground. They have nothing to do with Machu Picchu, or, for that matter, UFOs, another subject Berkland once brought up in a discussion on prediction, although the connection he was after remains unclear. At the same time, I'm sufficiently intrigued to give him a bit of leeway, and more than once, I've been astonished by the unexpected sense he makes. When I asked, for instance, what he'd learned there in the King's Chamber, he answered, "To seek your purpose and strive to achieve it. Anything else is a waste." It's hard to argue with these as words to live by, as it is with his insistence that "a scientific fact is at best a progress report."

This enigmatic quality has fascinated me about Berkland from the outset; it's the reason I've decided to go and meet him after all this time. It's hard, after all, to read someone from such a distance, hard to know exactly what he means. Since the late 1990s, I've periodically haunted Berkland's Web site, a looping electronic flash point that, in gathering his forecasts and those of others, seeks to operate as an information clearinghouse, although most often, I log off confused. I've read his monthly newsletter, *Syzygy*, which takes its name from a key component of his prediction theory and features letters, bits of personal history, a small photograph of the author, and a running commentary on his hits and misses for the year. I've talked to him, and talked about him, both to people who think that he's a charlatan and to those who suggest there might be merit in his ideas. Through it all, however, Berkland himself has remained elusive—ink on paper, binary code on a computer, a disembodied voice on the telephone. Once, after a particularly cryptic conversation, I opened up a copy of his newsletter on my computer, and sat for several minutes studying his picture on the screen. At regular size, it was not much larger than a postage stamp, an undistinguished portrait of an older man sporting a white beard and a floppy hat, grinning in a fixed and distant way. After a while, I decided to enlarge the image, but when I did, it broke down into black and white pixels, a field of dots and dashes I could not decode. I adjusted the contrast, as if this might help provide a cipher, but the closer I looked, it seemed, the more Berkland grew obscured.

In the early afternoon, I forgo the 5 for I-580, and ease out of the Central Valley towards San Francisco Bay. It's been a nondescript drive ever since I left the Grapevine—or perhaps it's just that, outside Southern California, I don't know the landmarks all that well. I do know that in 1983 the town of Coalinga was struck by a 6.5 that some seismologists think may have been the missing Parkfield earthquake; Coalinga, after all, is only twenty miles or so northeast of Parkfield, and there's little doubt the temblor there changed stress conditions along the fault. To me, however, Coalinga is most memorable as a cattle town (*Cow*-alinga, my Central California friends call it) that announces itself twenty or thirty miles down the freeway with the overwhelming scent of fresh manure, a smell so thick and loamy that it penetrates everything, even air-conditioning at full throttle, leaving my eyes red and teary and my throat parched and raw. Because of this, when I finally start running west along 580, it feels like a return of sorts, a reentry to the California I understand. I cut through the Altamont Pass, its hills lined with rows of large propeller windmills. I pass through Livermore, where my first earthquake may or may not have happened, and into Hayward, the town that has lent its name to the fault responsible for the second so-called Great San Francisco Earthquake, which struck on October 21, 1868, damaging structures, including San Francisco City Hall, on both sides of the bay. This is familiar territory, and I'm starting to feel comfortable, until I merge onto I-880, better known as the Nimitz Freeway, which holds an infamous place in California seismic lore for the collapse of its double-decked Cypress Street Viaduct in Oakland during Loma Prieta, a failure that killed forty-one people, two-thirds of the fatalities from the quake. Another October earthquake, I think, like 1865 and 1868, and this recalls something Jim Berkland once told me—that, in the Bay Area, October is a particularly treacherous month. The thought gives me a little tickle, a tremor of anticipation, but, as always in earthquake country, denial is as powerful a force as superstition, and the world around me looks so solid . . . I put the possibility out of my mind.

I cross the Dumbarton Bridge into Menlo Park, wind my way through shaded suburban streets to the Northern California field office of the USGS. Unlike its Southern California counterpart, this USGS office occupies a complex as sterile as an industrial park, a low-lying honeycomb of buildings spread around a parking lot, with a supplemental suite of offices out back. Although I've never been here, I've always considered Menlo Park a looser, more intuitive environment than Pasadena; it's where the Parkfield experiment got started, where it continues to be coordinated even now. Yet no sooner do I enter than I find myself longing for the casual informality of Linda Curtis, the cartoons tacked up in her office, the oddly comforting confusion of the X-Files, the way she functions on a human scale. Here, the halls are dotted with charts detailing twentieth-century earthquakes, but the effect is stifling, as if all that had happened at a remove—it's impossible to make the leap, intellectually or emotionally, from these abstract images to their impact in the world. Once again, I'm brought face to face with the inadequacies of science, its inability to evoke the mystery of earthquakes, to address the awe they make us feel. I stop before one map, trace the line of the fault against the vaguely three-dimensional uplift of the mountains, but what I mostly see is the image's flatness, its failure to penetrate the surface of the temblor, to capture, in any significant way, the depth of the event. It's like looking at that computer picture of Berkland, or even those old photographs of 1906 San Francisco: I feel suspended, out-of-time, caught between proximity and distance, as if the connection I want, the definition I'm seeking, is destined to remain always slightly out of reach.

My discomfort softens somewhat once I find my way to Allan Lindh's office, which is in the back building, around a corner and down a long corridor, like the hidden treasure in a scavenger hunt. Lindh, after all, is more what I expected—a fifty-eight-year-old geophysicist who believes in possibility, the idea that risk is in the nature of the field. Although he's worked at Menlo Park since the early 1970s, he is animated, unreserved, exuberant, a bantam of a man, thin and wiry, with frizzy blond hair that cascades to the middle of his back in a ponytail and a long bushy beard that he flips up and down with one hand as he speaks. His office reflects this energy almost perfectly: a twelve-by-twelve box cluttered with books and papers and electronic equipment, it looks like it contains the detritus of everything he's ever thought about, and stands in sharp contrast to the antiseptic stillness beyond this room. Yet even as I mark Lindh as an outsider, I'm struck by his self-deprecating, no-bullshit manner, as if, seismologically, he has seen and done it all. At first, this leaves me unsure how to read him, whether he's coming from the edges or the center, the extent to which he fits into the conservative culture of the USGS. Soon, however, I see that such contradictions do not matter—or, more accurately, that they're resolved by Lindh's ability to straddle the line between science and supposition, to loosen those polarities and realign them in a more inclusive point of view. Nearly a quarter of a century ago, in 1978, it was Lindh who pushed the USGS into investigating Parkfield, a decision he still defends despite the earthquake's failure to materialize. "We realized in the 1970s," he says, "that there's no point putting out instruments unless you're going to put enough in one place so that if something does happen, you'll record it on multiple instruments and gain some physical understanding of it. We knew it was taking a chance. We thought our odds were better than they turned out to be, but"—and here, he shrugs and gives a little chuckle, pale blue eyes glinting with mischief—"that's life."

Lindh's offhanded response to Parkfield is as unexpected as it is compelling, revealing his openness, his humor, his willingness to admit that he has made mistakes. The discredited effort to interpret the 1934 Parkfield earthquake as having arrived ten years early, for instance, was his idea—an error, he now admits, which together with an overreliance on anecdotal evidence from the nineteenth century, may have led the experiment "awry." In the next breath, though, Lindh tosses out a variety of other possibilities, invoking earthquakes from Coalinga all the way back to Fort Tejon, speculating as to their residual impact on the fault. When I ask about the public's sense of Parkfield as a washout, a wild goose chase, he laughs and acknowledges the perception, but insists that this is only one side of an issue far more intricate and complex. "The experiment," he says, "always had two parts. There was the attempt to guess when the earthquake would occur, the short-term phenomena, and then there was the effort to measure things very carefully and gather information, which people often overlook." In addition, Parkfield is important as the spot where "a great earthquake got started. That's the other reason we went there. It isn't just a place where they have sixes. It's a place where, once upon a time, they had a magnitude eight." The temblor to which Lindh is referring is, once again, Fort Tejon, which grows in stature, in significance, the more I travel on the California earthquake trail. The 1857 Parkfield earthquakes, in fact—there were two, both measuring about 6.0—appear to have been foreshocks of Fort Tejon, which followed them by only a few hours, at a distance of fifteen miles. "Their relationship to the big earthquake is unequivocal," Lindh says, and this, in turn, adds yet another layer to the work at Parkfield, since any event there might also be a foreshock, a warning of something larger still to come.

As Lindh speaks, his voice rises in excitement, and his laugh lines crinkle with enthusiasm behind his beard. It's infectious, especially when he discusses prediction, a subject about which he remains unapologetic, despite the sense of many in the scientific community that it is (to steal a phrase from Lucy Jones) "the kiss of death." If you ask why prediction evokes such strong emotions, Lindh will tell you it's a reaction to the collapse of theories like dilatancy-diffusion, which left many seismologists feeling disillusioned and betrayed. "You have to realize," he explains, "that initially, dilatancy-diffusion looked like a miracle, a magic signal coming out of the earth. Some people never recovered from the experience of having been seduced, in their youth or middle age, by something so wonderful. When it let them down, they were very much like jilted lovers, and were never willing to go there again." The irony, Lindh believes, is that, more than any fantasy of forecasting, it's such a backlash that is unscientific, a wholesale abdication of seismology's most essential goal. "To say that earthquake prediction is impossible," he declares, "is absolute nonsense. It reflects a failure of understanding, a failure of any kind of grounding in the philosophy of science. Science depends intrinsically on prediction. There is no science without prediction because you formulate hypotheses, those hypotheses make predictions, you then test those predictions, and they fail or succeed. If you can't predict things and then test them, you can't function in science in the most ordinary sense. To say we currently don't have the technology or the knowledge to predict earthquakes, that's clearly true. But to say we can't develop the knowledge, or we can't develop the technology, that's just crazy."

On the one hand, Lindh is absolutely right about the relationship between science and prediction—the two are intricately related, on as many levels as you can name. Still, sitting in his office, I'm struck by the subversive nature of what he's saying, which makes me wonder again about his place here, the extent to which, despite the longevity and influence of his career (he was for many years head scientist at Parkfield), he is an outsider, after all. Most important, though, is what his comments have to tell us about seismology, about the direction the field has taken and the ones it could have taken, the serendipitous ways that knowledge and assumption interact. From the standpoint of the present, it's tempting to look at history, any history, as fixed, inevitable, the record of a world where everything that has happened should have happened, and events have gone the only way they could. Yet Lindh is offering an alternative vision—a minority report, so to speak—in which we glimpse another possible way of seeing, a parallel seismological universe. That's not to suggest he is in any way cavalier when it comes to science; he was, remember, primarily responsible for debunking dilatancy-diffusion, with a paper Carl-Henry Geschwind has called "very careful and devastating"—this at the same time he was establishing his own prediction experiment at Parkfield. He is also exceedingly realistic about what prediction has to offer. "Of course," he wrote in "Can We Predict the Next Earthquake?" a 1990 article published in MIT's *Technology Review*,

> even if we instrument . . . fault segments, earthquakes will remain extremely hard to predict. Faults are enormously complex physical systems. The record of prior earthquakes is always too short, the data incomplete, and the models imperfect. We will probably never be able to forecast earthquakes with the same accuracy with which we can forecast, say, the weather. And inevitably, there will be false alarms. Still, trying to improve our ability to predict earthquakes brings discipline to the science of seismology and is important as part of a broad effort to heighten public awareness about earthquake hazards.

As thesis statements go, Lindh's is a cautious one; he's not talking about prediction in the same way as, say, Jim Berkland or Cloud Man, making what are (as of yet, at any rate) scientifically insupportable claims. What he's suggesting, actually, is not so far removed from the USGS's own strategy for hazard assessment, which relies, in part, on long-term probabilities, like the oft-repeated warning that there's a 67 percent chance of a 6.8 or larger earthquake hitting the Bay Area in the next thirty years. The difference is that while the USGS bases its evaluations on pure statistics, Lindh has in mind a more refined system, involving narrower geographic and temporal windows, and having less to do with numbers than instrumentation, data collection, direct observation of seismic phenomena, firsthand information gathered in the field. That, too, is nothing if not elemental science, the idea that there is no substitute for the physical work, for interacting with the environment, for having a long, hard look around. Earthquakes originate ten to fifteen kilometers below the surface—that's "ten to fifteen kilometers of mashed up rocks and water," Lindh notes with a rueful grin, "and it's forever being squeezed and shaken by other forces. It's a real mess." To have any hope of understanding them, you must sort out that mess, make sense of those complex forces, see the parameters of seismicity for what they are. "You need to know what's in the earth," Lindh says, "if you even want to *talk* about prediction." It is this, he argues, that continues to make Parkfield so important, because, if nothing else, the material collected there over the years adds significantly to our knowledge of that segment of the fault.

All these issues come together in the area of Loma Prieta, which Lindh has long characterized as a forecast event. Interestingly enough, he's not alone in that assessment, although "forecast" means different things to different people, depending on how they define their terms. In her 1998 paper *The Loma Prieta, California, Earthquake of October 17, 1989—Forecasts*, USGS seismologist Ruth A. Harris sets the standard this way: "An earthquake forecast is defined as a statement that an earthquake is expected to occur within a period of a few years to a few decades." She goes on to catalogue "18 studies published between 1910 and 1989 that variously offer or relate to scientific forecasts of the 1989 Loma Prieta Earthquake." For Harris, then, forecasting is a bit of an amorphous enterprise, encompassing everything from Harry Fielding Reid's early warnings about future temblors on the San Andreas to prognostications issued only a couple of months before the Loma Prieta quake. If you look at the work she's gathered, however, you can't help noticing a significant increase in the number and specificity of forecasts throughout the 1980s—fifteen of the eighteen reports she cites date from 1980 or later—including those made by, among others, Christopher Scholz and Lindh himself. "There are people," Scholz notes, "who would tell you that earthquake was forecast. It was forecast at the level of saying, 'Look, this place is ready to go; it's going to be about this big; it's going to be about here, and it's going to happen pretty soon.' You can't do any better than that." For his part, Lindh is just a bit more measured, saying simply, "It was a partial success."

No matter how you look at it—whether you agree with Scholz and Lindh, or take the opposing viewpoint, which reads claims like theirs as more consequential in hindsight—there's no question that Lindh was on the front end of the forecast curve. He first became interested in the Santa Cruz Mountains (where the Loma Prieta rupture would occur) in 1982, when he began to make calculations about that segment of the San Andreas, which paleoseismological evidence indicated had broken at least three times in the nineteenth century, in contrast to its dormancy since 1906. Such a state, Lindh believed, was deceptive; it masked a strain buildup that might well be accelerated because the 1906 earthquake had dissipated as it moved south from San Francisco, in response to certain obstacles in its path. These, Christopher Scholz would argue three years later, included a nine-degree curve in the fault beneath Black Mountain in Palo Alto. As Scholz explains it: "When the earthquake propagated south and hit that bend, it pumped its energy into the mountain, which produced shattering all over the mountain, but also caused it to run out of gas. Further up the Peninsula, the quake caused a slip of about three to three and a half meters, but once you get south of Black Mountain, the slip is only about a meter or a meter and a half. If you take that differential, and make a model of the strain, the kinetic data, you'd have to say that segment is ready to go. And you can tell how long it would be, and how much slip, which tells you what magnitude of an earthquake it would make."

Lindh's take on Loma Prieta is slightly different, although, like Scholz, he acknowledges the role of geography in the increase of strain along the fault. "Faults don't go in perfectly straight lines," he says. "They go around corners, they come to bends, weird things happen, big chunks of granite get in the way. Stress builds up. There's a big bend here, in the Santa Cruz Mountains, and the Loma Prieta earthquake basically took up more stress in the bend." Still, Lindh notes, if strain alone were the issue, there would have been no way to forecast Loma Prieta in any but the most general terms. Early in 1988, for instance, the Working Group on California Earthquake Probabilities, a panel of which both Lindh and Kerry Sieh were members, issued a report on the prospect of major earthquakes along the San Andreas. But while the Southern Santa Cruz Mountains were identified as the most likely locus for a Bay Area temblor, the statistical analysis was so broad—a 30 percent chance of a 6.5 or larger within the thirty-year period from 1988 to 2018—as to be irrelevant in any practical sense. Then, on June 27, 1988, a 5.3 in the Santa Cruz Mountains near Lake Elsman caught Lindh's attention, making him wonder if a larger event might be imminent on that section of the fault. His concern was heightened by Lake Elsman's location—near Los Gatos, at the northern end of the segment he had pinpointed six years before. "We had not had a magnitude five there since 1906 that we knew of," Lindh explains. "Certainly, we hadn't had one in sixty or seventy years. We knew there was something to it. It was a weird earthquake."

"A weird earthquake?" I ask. "What made it a weird earthquake?" It's the first time I've ever heard a geologist invoke this word to describe an earthquake, although when you get right down to it, what better word is there to use?

"For one thing," Lindh tells me, flipping his beard up, "there hadn't been any. The last good-sized quake to affect the area had been in 1906. That part of the fault had produced some threes and fours, but suddenly there was a five, and it was in a weird place. Also, it wasn't a strike-slip earthquake, it was primarily a thrust earthquake. But mostly, it was just so damn close to the north end of where some of us thought things were getting ready to go."

As his comments suggest, Lindh was not alone in his concern over the Lake Elsman earthquake; in the wake of the Working Group on California Earthquake Probabilities report, a rough consensus began to build about the possibility of further movement on the fault. "For the first time in San Francisco Bay region history," Ruth Harris writes, "the California Office of Emergency Services issued an official short-term advisory regarding a potential large earthquake. This advisory assigned a slightly increased likelihood of an $M = 6.5$ event on the Santa Cruz Mountains segment of the San Andreas fault in the next 5 days." Needless to say, this larger earthquake didn't happen, but within thirteen months, on August 8, 1989, a second temblor—this one a 5.4—shook Lake Elsman, in almost the exact same spot as the earlier quake. Again, the Office of Emergency Services declared a five-day advisory, and again, the anticipated mainshock did not appear. Two and a half months later, however, on the afternoon of October 17, Loma Prieta broke along a twenty-five-mile section of the San Andreas, extending roughly from Highway 17 near Los Gatos south to the Pajaro Gap. Measuring 7.1, it matched the magnitude range assigned by the members of the Working Group, and if you look at a map comparing the parameters of their forecast with the actual earthquake, you can't help noticing how closely the prospective and the real events line up. "We had two magnitude fives," Lindh says, "right at the north end of the segment within a year and a half of each other. Warnings went out, newspapers covered them, and if the damned earthquake had occurred in August 1989, five days after we issued those warnings, most of the world would have said that we'd predicted it. But it's probably just as well it didn't, because it would have been dumb luck."

Lindh laughs, as if to punctuate the irony of what he's saying. Here he is, an advocate of prediction, measuring what looks like a successful forecast in terms of luck. The truth is that he's right, though, or maybe it's that the jury is still out. There is, after all, a bit of luck involved in every successful prediction, and anyway, as with Haicheng, the more important question isn't what happened to anticipate this earthquake, but if it can be done again. "Look," Lindh says, voice rising in exasperation, or amazement. "It's not like the people who are still working on earthquake prediction are loonies. In fact, I don't know anybody currently working on it who is not extremely skeptical. They've learned this the hard way. But I do believe, and have believed for a long time, that there are good physical things you can observe that will tell you where you can expect a rupture. Good physical things you can see at the surface that suggest an earthquake's on its way." Such indicators, Lindh continues, encompass not only potential foreshock activity but also changes in rock density, variations in fault configuration, and topographical anomalies including fissures and other evidence of slippage on the surface of the fault. Still, all the indicators in the world won't add up to anything if we can't frame the necessary context, the necessary filter through which to read them right. Or, as Lindh tells me just before I leave his office, "The real question is how much information you can get, and there are only two ways of doing that.

One is to collect more, and the other is to understand better what you've already got. The day before they discovered the Rosetta stone, Egyptian hieroglyphs were gibberish. The day after they discovered the Rosetta stone, they were coherent sentences that described in detail thousands of years of history. They didn't have to make any new measurements, they just figured out what they had. That's what we haven't been able to do yet, to figure out what we have."

The day after my visit to Menlo Park, I see an earthquake cloud. It's late morning, around eleven thirty, and I'm driving slowly down a strip of two-lane blacktop in the small Sonoma County town of Glen Ellen, looking for the turnoff to Jim Berkland's house. His street is small, or so he's warned me, little more than a right-of-way, but although I have a list of landmarks to use as markers, I have the feeling that I've gone too far. I pull off to the shoulder and reread my directions—yes, I've crossed under the steel truss bridge, gone past the Sonoma Mission Inn; I've seen the horse farms and the wineries, driven by the Jack London Bookstore. It should be right here, *right here*, and yet I'm at a loss for where to go. I back the car up and turn around, deciding to retrace my steps, when all of a sudden, off in the north distance, a single cloud, thin and narrow, catches my attention, like a brushstroke overhead. I look at it more closely: it's elliptical, with scalloped edges, like the feather of an old quill pen. Once, at a café on the Caltech campus, I watched as Cloud Man pointed out the shapes in his now missing photos, trying to see what made them different, what set them apart. Yet in the end, his images were too distant, and while a couple took the form of feathers, and one did resemble a lantern, they mostly looked like clouds. How could he say for sure they were precursors? How did he know they weren't regular clouds? "Experience tells the difference," he told me, an elliptical response to an elliptical question, in which meaning was as elusive as, well, clouds. Today, however, I can see for the first time what Cloud Man meant, the way these clouds *are* different, unlike any I have seen before. I try to calculate the distances, try to figure out where it's pointing, but the rest of the sky is clear blue and without a context, and I can't tell anything at all. Maybe it's pointing me to Berkland's, I think, and smile thinly; maybe it's another sign. On the one hand, this is just a joke, a game of magical thinking, but on the other hand, I feel like I'm in mythic territory out here, and stranger things have happened, after all.

Strange things have been happening all day, in fact, or at least since I left San Francisco after breakfast and began to make my way north. Because I haven't come to Northern California in a while, I've forgotten about the pull of the place, the overlay of memory, the way that being here can sometimes feel like going through an excavation, like making an expedition into my past. Why this is, exactly, I can't tell you, unless it's that I lived here so briefly, and so long ago, that the experience seems stratified, locked in place beneath the surface, until a visit brings it back with seismic force. Either way, as I cross the Golden Gate Bridge and pass through the rainbow tunnel, I'm hit with wave upon wave of images, collapsing twenty-one years into a millisecond, like a human metaphor for geologic time. Entering Mill Valley, I think of Lauren's grandparents, wonder if they're still living, how their cantilevered cottage fared when Loma Prieta rolled through town. I remember my first earthquake, and the theory of Atlantis, which reminds me, for the second time in two days, of the futility of apocalypse, the fallacy that is the end of the world. Had the Atlantis story been accurate, after all, everything I see here would have disappeared nineteen years ago, erased in an instant, as if by the hand of God. It's a romantic concept, and I'd be lying if I said it didn't fill me with a kind of awe. The difference now, though, is that I no longer believe in finality, but rather process, chaos, ebb and flow. I think, in other words, that erasure operates in increments, that the Big One has no resolution, that Armageddon, as a moment, mostly never comes.

Of course, no sooner have I had this realization than the 101 sweeps into Novato, and I find myself driving up the exact embankment where my car rolled two decades ago. I haven't thought about this, haven't *allowed* myself to think about this, but as the freeway curves and rises, every bit of sinew in my body clenches in a muscular memory, and I can hear the sharp pop of the tire blowing, feel the jerks of movement as the world goes out of control. I slow down, try to calm my ragged breathing; "Take it easy," I say out loud. But although I coast through the turn, and, following Berkland's directions, get off at the next exit, it's as if something in my brain has shaken loose. First, I take the wrong fork at the Sear's Point junction, and end up on a long causeway headed for Vallejo, where there are no exits, nor even mile markers, only billboards for a seismic retrofitting company whose equipment lines both sides of the road. Once I turn around, I still can't find the route I need; I stop once, twice to ask for help before I get going the right way. Even in Glen Ellen, the most essential details still elude me, down to finding Berkland's home. Erasure, I decide, may operate in increments, but those increments can add up to something. This is how it is with earthquakes, the way they build and lock and strain for centuries, until all it takes is one last little bit of pressure to set them off.

I stare at the cloud for another minute, watch it hover, seemingly motionless, in the sky. Then, I put the car in gear and start to inch along the road, eyes wide open, as if I might will Berkland's right-of-way into being. I sweep down the street, return to the bridge, turn around, and take another pass. Just as I'm about to give up, I see it: a single paved lane running uphill, perpendicular to the road. I've been by this spot at least three times, but although a small neat sign is visible, I don't remember noticing it before. Maybe there *is* something to this magical thinking, I consider briefly, or maybe it's a matter of nuance, of being present, of paying attention to what you see. That's what Berkland would tell me, what he *has* told me; it's one of the arguments he uses to pillory mainstream science, which, he believes, ignores evidence that doesn't fit prevailing points of view. "How can you know what you're going to see," he asks, "before you see it?"—a question I can't keep from asking when I crest the small hill and turn left, into Berkland's driveway, which boasts a small sign reading "Berkland Way." My God, he's got his own street, I think, and all of a sudden I'm confronted on a whole new level by his enigmatic nature, his elusiveness, the way that, no matter what, he remains impossible to pin down. I park in front of a small outbuilding and make my way through tall grass past a fence to Berkland's house. It's a boxy structure, vaguely Victorian, with a pastel color scheme and detailed woodwork on the lintels. Hanging from the beams of the wide front porch is an enormous American flag.

As I'm trying to figure out what to make of this, the front door opens, and Berkland emerges to greet me, as if he's been waiting just inside. I recognize him from his picture, but there is something else, an intangible I can't quite put my finger on. At seventy-one, he's sturdy, barrel-chested, wearing white athletic socks, blue work pants, and a Jack London Historical Society T-shirt; his eyes are clear, pale blue, and piercing, with a hint of distance to them, the flicker of a thousand-yard stare. At first, I think this is it, these eyes his photograph has denied me, the way they reveal him, display his faraway nature, his tendency to be here and not be here all at once. As we start to talk, however, I realize that what's perplexing is how . . . well, *non*perplexing everything is. Standing on the porch, while Berkland points out where the original house—the house in which he grew up—sat on the property, I feel like I've entered a universe constructed entirely from images of a lost, traditional past. Here, after all, is a man who lives where he was raised: not just in the same town, but on the very piece of land. Later, he'll show me pictures of the old house, but for now, he talks about his dog, Minnie 2, a Border collie–coyote mix named for another dog he had that died. He is a devout churchgoer, a self-proclaimed patriot; his flag, he crows like some kind of proud uncle, once hung over the United States Capitol, and he'll carry it the following day in a parade at the Glen Ellen Fair. How, I wonder, does all this jibe with Machu Picchu, not to mention his predicted earthquakes—five hundred in the last twenty-eight years? The disconnect is so pervasive I can't find a point of reconciliation. Only at the last minute does his iconoclastic side assert itself, when he asks me to remove my shoes before we step inside. By the door, he points to a replica of a paving stone, on which is inscribed: "Syzygy Earthquake Newsletter. Editor Jim Berkland, Geologist, Predictor of the World Series Quake." The original, he tells me, is at Pacific Bell Park, in San Francisco, in front of the Willie Mays entrance. "I hope," he adds with a grim smile, "that my critics have to walk over it frequently, on their way in and out of the park."

Berkland's comment makes me laugh, although not for the reason he intends. To me, it's more a matter of irony. Of the predictors with whom I've spoken, none, including Charlotte King, stirs up such strong emotions; were all Berkland's critics assembled, they might well fill a good-sized ballpark by themselves. Just the sound of his name is enough to get some people going. Once, I mentioned him to Tom Heaton, only to watch Heaton's face cloud over and his normally easygoing demeanor disappear.

"Jim Berkland," Heaton muttered in a voice thick with annoyance, "has made many claims about his ability to predict earthquakes. He's made these claims for twenty years, and we've had lots of earthquakes. If the relationship between tides and earthquakes was that straightforward, it would be hard to suppress." To be fair, there's history between the two men—in the mid-1970s, as a Caltech graduate student, Heaton wrote a paper suggesting a possible correlation between tides and "shallow thrust earthquakes," but when he later retracted the theory because his research could not be duplicated, Berkland assailed these newer findings as dishonest, the result of pressure from the USGS. Yet even Allan Lindh, who has a good word for nearly everyone, reacts to Berkland with a certain attitude of disbelief. "Jim's a nice man," he says, "but he's been refuted in the scientific literature. Both the business about lost dogs and cats, and his simple-minded version of tides. Tom Heaton and Marsha McNutt, two of the smartest people ever to pull on a seismologist's boot, just ate his lunch on the cycles of the moon business, and some biologists at Davis did a good job on the dogs and cats."

Lindh is both right and wrong when it comes to Berkland; like most things in earthquake country, the truth is more complex, more elusive, occupying more of a middle ground. On the surface, after all, Berkland's theory operates from at least the semblance of a scientific premise—that the moon's gravitational pull can affect what are known as earth tides, especially during the "seismic window" that, he says, opens in the first six days after a full moon. Virtually all his predictions are bound by such a window, a period in which syzygy (when the sun, earth, and moon line up together at the full moon) and perigee (the moment the moon is nearest to the earth) can increase ocean tides from 20 to 100 percent. "When the earth is between the moon and sun," he explains, excitement striating his voice, "earth tides can create bulges of eighteen inches on either side of the planet. Once I realized that, I began to wonder if it was possible for the tides to stretch fault lines. Ground deformation can be a sign of impending earthquakes; it happened in Long Beach in 1933, and also in 1964 in Niigata, Japan." Certainly, earth tides are a documented geological phenomenon, and as recently as 1998, a paper presented at the American Geophysical Union meeting in San Francisco argued that lunar gravity could be a factor in about 1 percent of earthquakes—a figure that, while statistically negligible, does admit the possibility that something's going on. The same is true of animal anomalies, although Berkland's method of tracking them, which is to read lost and found columns in various California newspapers, is so anecdotal that there's no way to take it seriously, except in the most general terms.

This tendency to overgeneralize is what has always gotten Berkland into trouble, making him come off as grandiose, deluded, less a dreamer than a fool. The problem is his habit of drawing conclusions from circumstantial evidence, of hardening observations into rules. Take the origin of his prediction model, which he relates as soon as we sit down. "I arrived as county geologist in September 1973," he recalls, "and between then and December, I was greeted with six earthquakes in the Bay Area, ranging from 3.4 to 4.8. Come January eighth of 1974, an article appeared in the newspaper saying we might expect high tides along the seacoast due to an unusual astronomical alignment. What was this unusual astronomical alignment? Well, it was the first full moon of the year, and this was due only about an hour away from the moment of the closest approach of the moon in about five years, and within a week of the closest approach of the sun for the year. So those are the three major tidal forces—lining up the sun, moon, and earth, that's a syzygy; closest approach of the moon for the month, that's the perigee; and the closest approach of the earth to the sun for the year, which is the perihelion. When they happen almost simultaneously, you'll always have an extremely high tide. So I said, if the ocean waters are going up and down because of this arrangement, I wonder if the earth is also undulating. And if that's true, maybe it's stimulating fault lines into action. Let me see, we had these six earthquakes. One was the day after the full moon, another on the day of perigee; a third, two days after the new moon . . . all six fit this wild idea. They all happened within six days of the new or full moon. So I put it out to some of my other workers with the county: Hey, if this continues, we should have a quake of four or five magnitude here in the next week. I didn't take myself that seriously, except six out of six wasn't bad. But two days later, we got a 4.4. Right on the money. So I made another prediction for February, and that one also hit."

For Berkland, such a story represents a genesis saga, a creation myth, an account that, in the most profound sense imaginable, marks the onset of the world he now inhabits, a world that is nothing if not utterly mechanistic, as fixed and foreseeable as an enormous geologic clock. I, on the other hand, see it through a different filter, as a kind of cautionary tale. It's not that I doubt Berkland noticed something during those first few months in Santa Clara County, just as he did prior to Loma Prieta; the central question, though, is what. If you listen to him, he'll tell you it was a pattern, and not only that, but a pattern that fits almost every earthquake, which makes prediction, in his view, "routine." Yet while that may be a lovely fantasy, there's no real evidence to support it, since if temblors do occur in patterns, they are patterns that may not be measurable in human time. Certainly, this is one of the lessons of an experiment like Parkfield, that the earth is cryptic, intractable, and if the pattern there, which extended back nearly 150 years, was too short-term to pay off its prediction, what can we expect from an interval of months and weeks? Still, when I ask Berkland about it—gently, quietly, using the softest language I can muster—he doesn't even acknowledge that there's an issue, except to dismiss the possibility that he could be wrong. "It's a period of time when something is likely to happen, has a greater potential than normal, but it's not a certainty," he says, as if to explain away the vagaries of the seismic window. "And so it's similar to a weather forecast. But once you recognize a pattern, you should warn people. You see all these clues, and you put them together, and that helps build up probabilities."

Berkland's comment resonates like some inverse vibration, an antiecho of Lindh's take on prediction, that it will never be like weather forecasting, that the seismic record is too incomplete. This is the essential conundrum, the split between hard and soft science, and, as usual, I keep moving back and forth across that line. Although I agree with Lindh, I can't say Berkland sounds all that outrageous—until I recognize just how loose his terms are, how unresolved his probabilities, how amorphous their determination and their range. Even the patterns he claims to trace are more than a little slippery, especially in how closely predictions do or do not match actual events. In the September 2001 issue of *Syzygy*, for instance, he writes, "The strongest quake in the S.F. Bay Area within the August window was a 3.4M on August 21st. However, quakes of 3.5, 3.3, and 3.9M near San Juan Bautista were only a day late (August 25th). I would score myself at least 90% for those near hits." On the very next line, he offers a similar hedge in regard to Southern California, where "the strongest shaker in August was one of 4.4M near San Clemente Island on August 16th. Tantalizingly, this was also outside my window by one day, but I again would rate it as a 90% near hit."

I ask how he can lay claim to earthquakes that take place beyond the seismic windows, but he only slips into another digression, telling me about a 6.8 near the Salton Sea in 1979. For a minute, I'm not sure what he's getting at, until he mentions having seen, in a story about the temblor, a list of ten other quakes to strike the area since 1903. "I thought," Berkland says, "I'd see how they fit my windows. Not one of them did. They all hit from day eight to day thirteen after the new or full moon. My window closes on day six, usually. But I realized that the Imperial Valley has its own windows, which I wouldn't have guessed in advance. Why would that be? Well, in the Imperial Valley, we have geothermal wells. Very high heat flow at shallow depth. So apparently, it's more difficult to transmit strain through that kind of material. It requires greater lag time." The solution? To extend his windows to accommodate these anomalies, which, he announces proudly, "worked out beautifully. And it made sense."

Of course, making sense is a relative concept, especially when you're trafficking in conjecture about the world. Were Berkland still what he derisively calls a "mainstream seismologist," he'd have no choice but to admit the evidence doesn't fit his theory, much as Tom Heaton once did. But since he's not, he can do just the opposite, adjusting the data until it reflects what he needs. Certainly, it's compelling to watch someone spin a legend that accounts for everything, in which inconsistencies may be read as documentation, and illusions metamorphose into proof. Yet there's something disturbing about it also, an element of hysteria as opposed to truth. You can hear it in Berkland's voice as he gets rolling, words slightly manic, voice raspy with breathing, like he can't quite get enough air. From the Imperial Valley, he segues to the Mendocino Trench and Santa Barbara, where the last few good-sized quakes have arrived around the summer solstice, a coincidence he finds too significant to ignore. As is his habit, he seasons his speech with facts, statistics, precise dates and magnitudes; at the same time, he covers the low table before us with charts and papers, evidence to back up what he says. Among these documents are a set of binders featuring a daily log of pet disappearances in Los Angeles and the Bay Area, going back twenty-two years.

Although I've never doubted Berkland's obsessive nature, it's astonishing to see it laid out like this in three dimensions, and as he flips pages and points to various numbers, I try to imagine him, late at night, poring over newsprint at the kitchen table, inputting data in long, neat rows. At the same time, I find myself face to face once more with all his contradictions, his strange conflation of the bourgeois and the bizarre. Berkland's living room, after all, is as neat and nondescript as a hotel suite, carpeted and quiet, with baby blue walls and Muzak playing softly underneath our conversation like an ambient undertone. On the sofa beside me, a wool throw pillow bears the slogan "Old Friends Are the Best Friends," while across the room, on a round table near the kitchen, a gold chess set sits beneath a pyramid of glass. As we talk, his wife drifts in and out of focus like a specter, quietly reminding him of appointments or searching out documents he can't find. The experience leaves me unsure of where I'm sitting, uncomfortable with everything it implies. This, I suddenly understand, is why I've avoided meeting most predictors, as if, were I to mix with them too closely, the wonder they stir in me might dissolve into something more ambiguous, less resolved. In Berkland's case, the discomfort is intensified because, with the possible exception of Cloud Man, he's the one I've been most captivated by. What I mean is that, in some sense, I've spent the last few years identifying him as a kindred spirit, another sojourner out to navigate the wildness of California, another surveyor of the seismic divide.

Now, I'm staring at the flip side of that identification, the line where fact and fancy start to blur. The more Berkland goes on about earthquakes, the more I see that, in his eyes, this is not a matter of wildness, or even asking questions, but rather an issue of control. I remember Lucy Jones saying that prediction was less about seismology than psychology, and while I've often dismissed this point of view as narrow-minded, today I'm not so sure. For Berkland, after all, the lithic world appears as clear, as stark and orderly, as this living room; there are no gray areas, just a certainty that borders on the proprietary, as if all these earthquakes belonged to him alone. Late in the afternoon, while looking through neatly catalogued file boxes in his back office, he even invokes the Bible, as if it might be the final word. "This story," he says, "is not original with me. It's been around a long time. In fact, if you go to the Book of Matthew, you'll see that Christ is on the cross, and the sky darkens about three in the afternoon, there's a total eclipse, and then the ground shakes, and the Roman soldier says, 'Truly this must be the son of God.' Three days later, an aftershock rolls the stone away. Whether or not it occurred as related there, at least the order was correct."

For a moment, I'm stunned into silence, unsure of how to react to such a statement, or even what I've heard. It's not that Berkland has his information wrong; when I check him later against the New Testament, he's quoted the Gospel virtually word for word. No, what troubles me is the weight he puts upon the passage, the way that, for him anyway, it's more than just an anecdote. The Bible, to be sure, is full of earthquakes, including one in the Book of Acts that freed the disciple Paul and his companion Silas from a Macedonian prison after they'd been jailed for spiritually "cleansing" a female slave. This particular verse is iconic enough to have inspired the essay "My Dungeon Shook," with which James Baldwin opens *The Fire Next Time*. The difference is that while Baldwin frames his temblor as a metaphor, Berkland seems to regard *his* as a piece of literal truth. In that sense, what he's offering here is his own peculiar brand of seismic creationism, as if he were an earthquake fundamentalist and this his proof.

As for why this is, I get a hint towards the end of my visit, when Berkland admits that, during his early years as a geologist, "I would have been embarrassed to talk about prediction. I was sort of the fair-haired boy. I had just published several very important papers that Stanford professors were claiming would be the next big thing, and I didn't want to sully my reputation." Like most things in this indefinite universe, such a statement says a lot or nothing, depending on how you read it, and its meaning is utterly independent of its accuracy, since, at a distance of nearly forty years, who can say whether it's true or not anymore? What I can say, though, is that Berkland believes it, that he must see himself in these terms. If prediction was so dangerous to his reputation, then by embracing it, he has enacted an almost mythical fall from grace. He has allowed himself to become a martyr, dismissed, discredited, disregarded by a scientific establishment that once considered him its "fair-haired boy." That's a hell of a price, and in order to recoup, he has to be right—all the time, about everything. This is his redemption against having thrown it all away for a pipe dream, a lifetime of illusions, a slow dance of desire and self-deception out on the earthquake line.

And yet . . . the thing about redemption is that it often comes when you least expect it. Just as I think I've got Berkland figured out, he hands me the most recent issue of *Syzygy*, and directs my attention to the back page. After I turn the newsletter over, I notice a column, entitled "Aftershocks," offering Berkland's response to September 11, 2001, which he calls "perhaps the most horrific [day] in the history of mankind." I quickly skim the long opening paragraph, with its recap of the disaster, but I slow as I get to the bottom, where the following passage stops me cold. "On my website," Berkland has written,

> Daniel Perez of Michigan had posted on August 2, 2001, about his "premonition dream" of an earthquake hitting New York. He "saw" Katie Couric on the TODAY show, but the TV signal blacked out. He wrote, *"She was talking about 10,000 people dying in a building. Either she was talking about the Empire State Building or the World Trade Center. In the dream the quake was very devastating because she could barely hold her emotions together on the air, and in the background of the TV image, I could see huge open spaces where tall skyscrapers once stood, and one tall skyscraper was billowing smoke from all its broken windows. While Katie was interviewing some survivors, more aftershocks hit the city and knocked out the TV transmissions once more. I usually don't write about dreams and I usually don't dream about far-away places, but this dream stuck out because it scared the 'you know what' out of me."*

"That's something, isn't it?" Berkland says. For the first time since I arrived here, even he seems to be at a loss for words.

It's hot when I get back to San Francisco, shimmering mirage heat, dry and baking, like all the moisture has been sucked from the air. "Earthquake weather," my friend Elaine laughs, as we sit in her backyard. "It happens three or four times a year. It was like this in 1989, before the quake. Everything was so flat, so still." I nod, but don't say anything, just marvel at how earthquakes affect us, even after all these years. If you've gone through one—as Elaine did with Loma Prieta, or me with Northridge—the experience is always with you, a memory that never fades.

Elaine's house is small and open, with two tidy bedrooms and an expansive eat-in kitchen, which leads out to a redwood deck. This afternoon, she's preparing for a modest cocktail party, and after a while, she goes back to cutting vegetables and icing up the beer and wine. It's just as well, for ever since I left Berkland's, I've been feeling disconnected, buffeted by circumstance, caught up in the idea that, no matter what I think I can say about earthquakes, there will always be a twist, a turn, a wild card, reminding me of all I do not know. Even here, drifting at the edges of Elaine's yard, I'm in the grip of something; although I hear the trill of a starling, the low, throaty blat of a crow—no earthquake, I tell myself, no earthquake for at least three hours—this does not reassure me as it usually does. It's not that I expect a quake to happen, not literally anyway. No, it's more like one has come and gone already, and I'm living in that odd half-light of aftermath, a glow in which the normal edges of the world do not cohere. It was like this after Northridge, this sense that the most basic things (a tree, a leaf of grass, a window) were unfamiliar, that existence had slipped its bounds. In all the years I've lived in California, I've been both frightened and exhilarated by such a sensation, but this afternoon, my reaction is much more vague, amorphous, more of an ineluctable unease.

What it all comes back to is that dream, that premonition, that image of ten thousand dead in the World Trade Center, posted a month before on Berkland's site. No matter how much I go back and forth, I can't figure out what to make of it, how to contextualize it, how to *explain*. What does it mean if the World Trade Center was predicted? What does it mean if someone dreamt the Towers' fall? On the one hand, it's not so much of a stretch, given all the books and movies that deal with terrorist attacks, the computer games in which such scenarios get played out every day. "The city, for the first time in its long history, is destructible,"

E. B. White wrote as long ago as 1949. "A single flight of planes no bigger than a wedge of geese can quickly end this island fantasy, burn the towers, crumble the bridges, turn the underground passages into lethal chambers, cremate the millions. The intimation of mortality is part of New York now: in the sound of jets overhead, in the black headlines of the latest edition." *This* dream, however, feels somehow different, a matter of individual vision as opposed to Jungian archetype, which makes me wonder again if there are forces that exist beyond us, vibrations just below the range of human sensitivity, below the surface of the ground. The simplest interpretation would be to write it off as random, but as Allan Lindh told me yesterday, "There's no such thing as random. It's just a question of how much information you have."

All of that is only heightened by my confusion over Berkland, by the closed circle of his thinking, the way he frames his world in black and white. It makes me want to walk away from him, but the dream won't let me, the dream leaves everything unresolved. I can't stop wrestling with the dilemma, although I make an effort once Elaine's party starts and her friends gather on the deck. They're a small group, fairly intimate—ten or twelve people who have known one another, in some cases, for more than a decade, sipping drinks and sharing stories, catching up on the minutiae of their lives. For a while, it's a relief to listen, to sit and smile and not say much, to smell the crisp tang of eucalyptus, watch the afternoon stretch into evening, see the sky fade and grow mottled, shot with streaks of purple and pink. Eventually, though, the present can no longer hold me, and as the party continues, I begin to drift. How, I want to know, can I accept one thing and not the other? How can I make sense of ideas, of observations, which seem so inherently opposed? These, of course, are the same questions I've been asking from the beginning, but tonight, balance is more difficult to find. Is it possible that Berkland is both onto something and not onto something? Or is this all just a projection of my mind?

Finally, in the waning twilight, people start to get up, one by one, to shake out their limbs and move, in fits and starts, back inside. For the last half hour or so, everyone's been talking about *Ciudad de los Milagros*, an annual art exhibition that commemorates the Day of the Dead. Tonight marks the show's opening, at a converted warehouse space in SoMa, near downtown. When I ask Elaine about it, she describes it as a metropolis of space-specific sculpture, of installations and homemade shrines that together make up the pieces of an elaborate collective mind. The idea is to walk through, to immerse yourself, and, in the process, see what associations arise. "Sounds like what I've been doing all weekend," I say, half to myself, but the prospect, I must admit, is intriguing, and I'm eager for a change of scene. Maybe in the midst of all those installations, I'll find my own bit of redemption, or maybe I'll just stop asking questions for a while.

A bunch of us squeeze into Elaine's pickup and make the quick drive to the SomArts Cultural Center, on the corner of Ninth and Brannan, an old industrial neighborhood. From the outside, SomArts doesn't look like much, just a blank brick wall and an iron door, with a small crowd of people standing outside smoking, enjoying the tepid San Francisco night. "I'll look for you if you get lost," Elaine says, her hand on the doorknob, and I have exactly half a second to wonder what she means before she opens the portal and we are sucked into another world. Immediately, I'm assaulted by a cacophony of noise—voices, conversation, distant snatches of incidental music—then swept up in a crush of people, faces and bodies pressed against me, everyone trying to get in or out. All I know for sure is that I'm in some kind of labyrinth, with walls made out of white muslin stretched into a winding tunnel, which leads me deeper, deeper, into the belly of the beast. Every few yards, I see an alcove, or an entrance to some impromptu chamber, and everywhere are . . . altars is the only way I can describe them, altars to memorialize the dead. Around one corner, I discover a piece called "A La Memorial 11 de Septiembre," which features a small repeating waterfall that flows endlessly over a life-size image of a bald man with hands splayed out across his hips. A little further on, a skeleton in a white robe (Santa Muerte, a sign identifies her) stands as if a member of the audience, the scales of justice suspended from her right palm. I look around for Elaine, or any of the others, but they are nowhere to be found. No, for now at least, this City of the Dead is a city in which I know no one, a city where, much like the territory of earthquakes, I must navigate on my own.

I pause before Santa Muerte for a moment, then, as per the accompanying instructions, offer a novena for the safety of my wife and children, inscribing their names on small slips of paper, and leaving them atop the hanging scales. It's not the kind of thing I'd do, generally—I'm too self-conscious, too suspicious of religion, not to mention that, for me, the whole idea of novenas is as exotic as a set of hieroglyphs, indecipherable and obscure. Tonight, however, I find myself consumed by the notion that there's more at stake here—something deeper, more elemental, some form of shared experience I can't quite see. Partly, this has to do with Berkland, and that dream about the Towers, which continues to linger with me, like a signifier of some kind. In another way, it's what I've been looking for all weekend, a sense of order, of solidity, a measure of stability (or, at least, assurance) in a chaotic universe. Again, I'm reminded of that Easter Sunday with Lauren's grandparents, also a day of dislocation and strange juxtapositions, in which I found an unexpected equilibrium all the same. It's a long way from Mill Valley to

SoMa, as long as the twenty-one years it's taken me to get from there to here. Still, if I look at it right, I can see a certain way of thinking about the world. What I'm after is an intuitive landscape, a place where truth and imagination coincide. I want to be transported by uncertainty, put in touch with the ineffable, while at the same time keeping my feet firmly on the ground. I want to occupy a territory somewhere between Lindh and Berkland, in which geopoetry opens up a passage to both understanding and wonder, and earthquakes make us larger than ourselves. I want, in other words, to see the bigger picture, a picture that includes both faith and knowledge, a picture that has room for everything, that is, itself, the expression of a collective mind.

The irony, of course, is that such a vision describes almost perfectly the space in which I'm standing, much as Elaine suggested it would. What is this event about, after all, if not juxtaposition, intuition, the way we take the evidence we're given and shape it into a narrative with its own internal sense? It's a matter of making meaning, of taking our most shadowy intimations and finding their expression in the world. This emerges even in the exhibition's labyrinthine layout; although *Ciudad de los Milagros*'s planners may not know it, the labyrinth has represented throughout history "an archetype, a divine imprint," a path of discovery, of meditation, found in nearly every spiritual tradition on earth. There is a labyrinth, for instance, set into the floor at Chartres Cathedral, designed as a "prayer walk" for thirteenth-century pilgrims who could not make the journey to Jerusalem. Two years ago, in Santa Cruz, I walked a different labyrinth with a man named Bob Haden, for whom the maze, and our interaction with it, represented the serendipity, the unpredictability, of daily life. In Haden's view, the key was to recognize that every action, no matter how small or insignificant, had resonance, that everything connected to something else. "If you want to pass someone and don't," he told me, "that's a metaphor. If you get restless, that's a metaphor. If you get lost, that's a metaphor. The labyrinth is a way of metaphor, and if you let yourself go, you'll find metaphors for your life."

Haden's thoughts have long provoked in me a certain sympathy; I've always been a metaphorical thinker, after all. Tonight, however, as I walk *this* labyrinth, I feel them as an almost three-dimensional filter, especially when I turn a final corner and come upon a shrine I haven't seen before. Unlike many of the other installations, this one is unadorned and simple, a small table bearing a blank book called the "Book of Miracles," amid a backdrop of four rectangular projection screens. There are no saints here, no icons, just a sign asking visitors to transcribe their experience of the miraculous, which makes this the manifestation of a do-it-yourself sainthood, an altar where we sanctify ourselves. As I stand there, I notice a series of messages on the screens before me, but I don't really pay attention; I'm more interested in the book, in the idea of writing miracles, as if the act of setting them down on paper could be the first step in making them come true. Although, like the novenas, this is not the kind of exercise in which I generally participate, tonight I take up the journal and thumb through it without hesitation, looking for a sign. Mostly, what I find are commonplaces—"Thanks for the miracle of life," reads the most recent entry, from a woman named Darlene—which makes me wonder whether miracles are more ordinary than we imagine, or if we've lost the ability to appreciate them at all. Then, I turn to a blank page and begin to write about the dream of the Twin Towers, suddenly convinced that, if this is not a miracle, nothing is.

It's late when I finish, and the crowd is thinning; there is less energy, more silence, a sense that this has, indeed, become a city of the dead. I lay the "Book of Miracles" back on the table, and as I step away, I glance again at those projection screens, with their four messages that I have yet to read. And I want to be careful about the claims I make, the inferences I ascribe, but I also want to tell you what I see. And I know these statements have been on display all evening, that they are part of the installation, but I can't help feeling like they've been put there just for me. It could be coincidence, or it could be metaphor, or perhaps it's something in between. But left to right, one screen to the next, I read:

A scientific law is a generalization based on observation.

1) *Scientific understanding is always based on constant repetition of events.*
2) *Miracles are not constantly repeated.*
3) *Therefore, there is no scientific way to understand miracles.*

BEFORE AND AFTER SCIENCE

My favorite photograph from the Northridge earthquake is not an image of the devastation. Instead, it's a quiet, almost pastoral shot, taken on a San Fernando Valley street during what looks like one of the first mornings after the quake. It's a clear day, and in the background, you can see a verdant spray of vegetation, behind which sits a small brown house, utterly intact. To the right, a man in black slacks and a blue sports shirt walks purposefully along the sidewalk, a clutch of papers in his hand. The scene is so still, so quietly unstudied, that it can be hard to recognize the context, hard to see the moment for what it is. Hard, that is, except for one final element, a large sheet of plywood that takes up nearly the whole left side of the picture, on which is spray-painted in black and red block letters:

WELCOME TO L.A.
SOME ASSEMBLY REQUIRED

What I like about this picture is its sense of humor, the way it relates a kind of fatalism in the face of fear. I like its toughness, its attitude, its human take on geological phenomena, its ability to encompass both the short and long view all at once. Such a photo reveals something about how we interact with earthquakes—not as matters of science, but as matters of life. It's this same impulse that led World Series fans at Candlestick Park to hold up placards on which they'd scrawled projected magnitudes in the first few minutes after Loma Prieta; "6.0," one read, "Now let's play ball." If you haven't spent much time in California, that might sound a lot like denial, although, as a Los Angeles County emergency services worker once told me, denial can go a long way in earthquake country—without it, we might never get out from underneath disaster's weight. To me, though, these reactions suggest something far more significant, a quality of resourcefulness, of imagination, a resilience that enables us to make our peace with seismicity, and, in so doing, approach a state of grace. This is the sensibility that, in 1906, helped rebuild San Francisco, the notion that if we laugh at earthquakes, or even coolly disregard them, then in the most basic manner we may begin to control them, looking past their physical damage to balance their psychic impact on our lives.

I've been thinking a lot about balance ever since I returned from Northern California, trying to frame at least the semblance of a consistent point of view. When it comes to seismicity, that can be a tricky business, as my experience in the Bay Area goes to show. After all, if meeting Lindh and Berkland has solidified anything, it's my sense that earthquakes, by their nature, occupy a constantly shifting landscape, in which no skill is more important than the ability to hold opposing concepts in our minds. At heart, this is—*has* to be—an individual process, an issue of interpretation, of rewriting situations almost instantly, depending on the circumstances that evolve. Everyone who's ever been through an earthquake has a story, an idea of what it means. For me, Northridge is a tale of reconciliation; although I'd lived in L.A. for two and a half years before the temblor, I had never seen the city as anything other than ephemeral, a movie set construction that looked like it would crumble in the first strong seismic jolt. Then the earthquake came, and nearly everything I knew remained intact. Such a narrative is different from, say, that of my friend Harlan, who was thrown down a flight of stairs by the tremor; he had just woken up and was going to the kitchen to make breakfast, only to end up with a dislocated shoulder and a broken nose. There is a way, however, in which our stories move beyond the personal to connect at the level of the universal, because they arise out of a shared event. The same is true of that Northridge photo: while it may capture the reaction of a single person, someone else had to take the picture, and in the act of seeing, millions more have somehow integrated it into their experiences of the quake.

Does that—the conflation of interior and exterior, the individual and the collective—sound like the stuff of contradiction? It should hardly be surprising, for earthquakes are contrary, confounding, operating in all sorts of directions all at once. In many ways, they're not so different from *Ciudad de los Milagros*, which may be one reason the exhibit moved me, with its unexpected turns and strange connections, its sense of tapping into something just beyond my grasp. Simply to think about earthquakes is to discover yourself in a territory defined by paradox, where everything gets shaken up. Ambiguities pile up like aftershocks, ranging from the obvious—the relationship of past and present, eternity and erasure, the boundless depths of geology and the fleeting breath of human time—to other, subtler distinctions that get at the very essence of how we see the world. What, for instance, can we say about science and prediction? Is it possible, as Lindh suggests, that we may one day find a resolution, or are they wholly different paths? Since the moment I first became interested in earthquakes, I've felt suspended in this polarity, unable to come to terms with it but unwilling to turn aside. Rather, I find its traces everywhere, from Pasadena to Glen Ellen, not to mention all the epicenters along the way. In such a universe, even a photograph can become a contradiction, like a window opened by the excavation of some ancient fault. Each time I look, I feel that pull between memory and anticipation, between how things were, how things are, and how things will be again.

Faced with all these layers, I can't help feeling a bit of mental vertigo, as if I were trapped in an intellectual Möbius strip. It's hard to keep my equilibrium, to know where I am standing, hard to know what I expect from earthquakes, what, exactly, I am looking for. I want to know the facts, but then, with earthquakes, facts dissolve into contradiction also—or, perhaps more accurately, into fallacy. I don't mean facts are irrelevant, just that they're slippery, elusive, fragments of information we can use to support almost anything if we frame them properly. Of the people I've met on the earthquake trail, none invokes facts as much as Jim Berkland; they dot his conversation like signposts, bestowing what seems a measure of credibility, evidence that he really does know his way around. If we line up his facts, though, and attempt to navigate a passage, we end up with a map that leads us nowhere except back to ourselves. Again, the issue is interpretation; it's one thing to observe phenomena, to track magnitudes and dates in an effort to see patterns, and another to say what they mean. The difference is between gathering the pieces of a puzzle and putting the puzzle together, which becomes far more complicated with earthquakes, since the pieces are not always pieces we can see. That's true not only of Berkland but also of mainstream science, much of which, Tom Heaton notes, relies on "repeating a particular experiment or developing a theory. Unfortunately, the earth is not really susceptible to those techniques. We can't repeat an earthquake. Theories are difficult to come up with because we don't know all the conditions, exactly where things are happening, so we have to stand back and infer what's going on from an observational point of view." In a field where knowledge is so piecemeal, facts yield almost seamlessly to conjecture, which means a little information can be a dangerous thing. "You'll never get a scientist to say that information is detrimental," Lucy Jones once told me, "but we could do a better job of explaining how we use it—and we should."

Jones, of course, has always been a pragmatic figure on the earthquake landscape. In 1995, she, in association with the USGS and the Southern California Earthquake Center, wrote a pamphlet called *Putting Down Roots in Earthquake Country* that considered the information question in practical terms. Six years and a hundred thousand earthquakes later, her work remains exceptional for its directness, its ability to address the subject point by point, in clear, intelligent language that summarizes the state of science on matters ranging from the cause of earthquakes to their frequency and mitigation, before closing with a discussion of seismic safety, which she goes out of her way to stress. Yet while Jones is scrupulous in her presentation—she even debunks eight "earthquake myths," including earthquake weather ("Every region of the world has a story about earthquake weather, but the type of weather is whatever they had for their most memorable earthquake") and prediction ("Neither Caltech, its scientists, nor the scientists of any other research organization in southern California have ever successfully predicted an earthquake's time within days, nor do they know how or expect to know how any time in the foreseeable future")—something is missing, some intangible that facts alone cannot provide. I'm not sure how to explain it, except to say that facts and truth are not the same. Here are some facts I know about earthquakes: I know there is a "creep zone" on the San Andreas north of Parkfield, a hundred-mile stretch along which the fault slips without apparent friction, the result of a nearly constant run of microscopic earthquakes, far too small for us to feel. I know that my wood-frame house is an especially sturdy structure that, barring unforeseen circumstances, should ride out any temblor I'm likely to see. I know that the longer an earthquake lasts, the higher its magnitude, that if you experience a gentle tremor of long duration, it's really a significant shaker far away. Still, for all that this may be useful in pinpointing, say, a 6.1 in Orange County, it's ultimately little more than a parlor trick. What none of these facts offers is a point of view, a perspective, a strategy for tying things together, a way of looking at the forest, not the trees.

This is why so much science doesn't move us, why it has trouble capturing our imaginations, why many people turn away. This is why we look for myths, for markers, for enlargement, for an explanation that can do justice to our terror and our awe. Lucy Jones may be right when she says prediction is in the eye of the beholder, but there's another matter she's overlooking, which has to do with our desire to believe. Yes, it's an issue of control, as she suggests, but it's also about making sense of the senseless, finding meaning in the tumult that lurks beneath the surface of the everyday. Although no one keeps statistics on the number of people who listen to predictors, a simple Google search yields

14,200 potential hits for "earthquake prediction"; Berkland, meanwhile, claims to have recorded a million visitors to his Web site (as always, I have doubts about his numbers), and Cloud Man's online guest book is thirty-three pages long. It's hard for me to assess any of this objectively, since I straddle both sides of the line here, identifying with the longing while remaining skeptical of the results. Particularly in the wake of my visit to Glen Ellen, I can't really buy into most of Berkland's theories, because I've seen how he fudges numbers, how everything is about being right. At the same time, I have to give him credit for the accuracy of at least some predictions, and I can't deny having felt the same seductive weightlessness, the wonder he refers to when an earthquake seems to fit a pattern, even (or especially) if it's a pattern that I can't quite see. Once, at Caltech's Earthquake Exhibit Center, I spent an hour scrolling through a few weeks of Southern California earthquakes, trying to correlate forecasts to actual events. At the time, we were in the middle of one of Berkland's seismic windows, but despite a cluster of 3.5s and 3.8s along the California-Nevada border, nothing really fit. Those quakes, however, did seem to jibe with a forecast made by Cloud Man for a 4.5 or larger—although not completely, because the magnitudes were so far off. What was I looking at? I asked myself. Was there something to this, or was it all coincidence? That amorphousness, that floating quality, brings me face to face with yet another contradiction, perhaps the most fundamental contradiction, between what we see and what we feel. Where do we find a point of reconciliation? Does prediction tell us anything? And what about science? Can it give us what we're after, or are we looking at it the wrong way?

I find the inkling of an answer one night while having dinner on the terrace of a Mexican restaurant in Santa Monica, which is the last place I'd have thought to look. Or not an answer, necessarily, but another way of seeing, an approach that casts my confusion in an entirely different light. It's a cool autumn evening, and I'm with the science writer K. C. Cole, who's made a career of explaining the ineffable, of taking the vagaries of physics and framing them in black and white. As we eat, Cole talks about the beauty of mathematics, the complexity of logic, the way that in the area of physics, unknown (and possibly unknowable) factors almost always are involved. I've come here because I'm interested in Cole's ideas on prediction, which, as an element of science anyway, she considers largely beside the point. "A scientific prediction is less like a weather forecast than a train of thought," she explains in *The Universe and the Teacup: The Mathematics of Truth and Beauty*. "Predictions are used to test which theories are on the right track, going in the right direction. If the theory's predictions prove false, then it's clear the thinking has to change directions. Predictions are guideposts along the way to understanding, not goalposts." Obviously, this perspective is antithetical to the approach of someone like Berkland. But tellingly, it's also distinct from Lindh's view on the subject, his sense of information and expectation, the subtle ways they interact. When I mention his belief that there's no such thing as random, that it's more a matter of what we do and do not know yet, Cole cracks a gentle smile, and slowly shakes her head. "Maybe that's true," she says, "but the real issue is, can we ever know enough?"

On the surface, Cole's question seems an exercise in philosophical musing, open-ended, rhetorical, abstract. The deeper I look, however, the more it starts to sound like a statement of principle, pointing down a whole new path. Surely, information is important to prediction, as even John J. Joyce or Donald Dowdy would agree. Still, Cole means to tell us, information can be a dead end also, because there are so many factors at work in even the most simple things. For proof, we need only look at the weather, a metaphor that keeps arising on both sides of the prediction line. To Berkland or Lindh, weather forecasting is either a model for what earthquake prediction could become or a model for what it will never become, but according to Cole, the model itself may be inherently flawed. Weather, after all, is notoriously fickle. Despite our best efforts, it often confounds forecasts, operating more like a random system than one that follows certain laws. The reason, Cole writes, is a process called "exponential instability," which means that all it takes is a single variable ("for example, 'if someone lights a candle' ") to throw any progression off. This is the essence of chaos theory, that even the smallest changes can have vast, unpredictable results. The best-known example of that is the so-called butterfly effect, first proposed by MIT's Edward Lorenz, who, in the early 1960s, noticed anomalies in what should have been identical weather simulations, and set out to discover why. What he came up with, notes James Gleick in *Chaos: Making a New Science,* was the realization that "for small pieces of weather—and to a global forecaster, small can mean thunderstorms and blizzards—any prediction deteriorates rapidly. Errors and uncertainties multiply, cascading upward through a chain of turbulent features, from dust devils and squalls up to continent-sized eddies that only satellites can see." Or, as Lorenz himself put it, in the subtitle of a 1979 paper: "Does the Flap of a Butterfly's Wings in Brazil Set Off a Tornado in Texas?"

What does all this have to do with earthquakes? Like weather, seismicity arises out of a complicated mix of stresses, strains, and other processes—what complexity theorist James Crutchfield has called a "web of causal influences [that] can become so tangled that the resulting pattern of behavior becomes quite random." There are reasons, in other words, for how and why faults slip, but as to what this means for a given earthquake . . . those possibilities are so intricate, so susceptible to influence and interaction, that they multiply geometrically until it's impossible to keep track of them anymore. This, in many ways, is what Kerry Sieh is getting at when he talks about the opposing forces that emerge throughout paleoseismology, the subtle interplay between chance and order, human and geologic time. Once again, we're in the territory of indeterminacy, where it may be all we can do to see the big picture, to identify the larger patterns, without worrying too much about what goes on in between. "It's the same thing with breaking a pencil," Lucy Jones says. "If I take a pencil and I snap it, I know it's going to break. And I can tell you approximately where it's going to break because of where I'm pressing on it. But I can't tell you which microsecond it's going to break in. That's the problem with the geologic timescale. When we want to know which year the earthquake will happen, it's the equivalent of which microsecond the pencil's going to break. And then the question is . . . each earthquake is not breaking the pencil through. It's one more little crack forming within it as you deform the whole thing. To look at it another way, it's like you're pulling taffy. Exactly where is each little crack going to form as you pull the taffy?"

Questions like this have always struck me as, by turns, fascinating and frustrating, for the precise reason that they can't be answered, but keep going around and around. On the one hand, I'm intrigued by their elaborately self-referential logic, the way that "in complex systems," as Cole notes, "feedback loops connect the parts in such ways that one part affects the next, which in turn affects all the others, and so on." That, I want to think, is how earthquakes work also, as elements of some endlessly linked chain of events that resets and repeats across eternity, with neither a beginning nor an end. At the same time, I can't help recoiling against the tangled irresolution of such a process, which feels like something of a Gordian knot. It's why, despite my skepticism, I continue to have sympathy for the predictors, for their motivation if not necessarily their theories, their conviction (hope? belief?) that they can trace a set of underlying connections and, in so doing, map a world where everything, finally, measures up. Tonight, though, for all my back and forth on the subject, I feel I've caught a glimpse of something, some small, stray glimmer of coherence, a strategy of integration at the core. The key is an offhand statement Cole makes just before we leave the restaurant, as we're gathering our jackets and paying the bill. We've been discussing this very issue, the paradox of systems that can be both causal and chaotic, and what this may tell us about the nature of understanding itself. For Cole, it's another open-ended question, an emblem of the universe's complexity, the way it constantly confounds our best attempts to pin it down. "Even in a perfectly closed system," Cole says, "there may be limits on what we can actually know." This, she goes on, is the assumption behind a lot of theoretical physics, beginning with the Heisenberg uncertainty principle.

Cole's comment sticks with me the whole way home, like a burr lodged in my brain. It's less a thought than a hint, and initially, I can't see where it leads. The Heisenberg uncertainty principle describes the relationship between the position and movement of subatomic particles, and the impossibility of simultaneously determining both. First proposed in 1927 by the physicist Werner Heisenberg, it theorizes that "the more precisely the position is determined, the less precisely the momentum is known in this instant, and vice versa." Taken at face value, that's a very specific dynamic, in which seismology doesn't figure at all. But the thing about science, or so I'm learning, is that very little operates at face value, which means that ideas (or some ideas, anyway) can be applied across the board. In this case,

Heisenberg's belief that one piece of information precludes the other—that you can't know everything at once—could easily apply to almost any area of inquiry, from the elliptical half-truths of prediction to the processes of the earth itself. "Alas," Cole writes in *The Universe and the Teacup*,

> even perfect understanding doesn't always allow us to predict the future. Nature, it seems, has contrived to make even simple pattern perception unreliable under many common conditions. For example, no amount of understanding of the behavior patterns of atoms allows you to predict just where an atomic particle will be at a certain time. The best you can get is a probability. Or if you do pin down the particle's position precisely, you can't say anything very precise about its velocity. One can't be known without sacrificing the other.

To put it another way, the problem is one of observation, a realization that falls even more into focus after I talk to my friend Leelila, a former physics major, who gives me the contextual spin I'm looking for. "The way it works," she says, "is that there are certain things you can't observe because in the act of observing them, you change how they behave."

Leelila goes on for a few more minutes, but I'm no longer listening, for what she says has stopped me cold. It's not that I don't believe her; no, it's just the opposite, that I always have. What she's describing, the notion that our perceptions of the universe are unavoidably subjective, is among the central tenets of quantum mechanics, a discipline with which I had a brief but intense infatuation in college, although I haven't given it much thought in years. Back then, I used to love to lose myself in its abstractions, the way that, unlike the closed cycle of feedback loops, it seemed to speak of endless possibilities, of imagination as an elemental, universal force. Now, I find myself wondering if quantum physics might offer a passage to a whole new set of associations, which have less to do with theoretical fancies than the realities of the physical world. To get a sense of how this might work, we need look no further than the paradox of Schrödinger's cat, a thought experiment proposed in 1935 by Erwin Schrödinger, who two years earlier had won the Nobel Prize. In a 1980 article for *Scientific American*, Harold J. Morowitz explains the concept this way:

> In a hypothetical formulation, a kitten is put in a closed box with a jar of poison and a triphammer poised to smash the jar. The hammer is activated by a counter that records random events, such as radioactive decay. The experiment lasts just long enough for there to be a probability of one-half that the hammer will be released. Quantum mechanics represents the system mathematically by the sum of a live-cat and a dead-cat function, each with a probability of one-half. The question is whether the act of looking (the measurement) kills or saves the cat, since before the experimenter looks in the box both solutions are equally likely.

The purpose of the experiment, then, is to highlight what Morowitz calls "a deep conceptual difficulty"—the role of the observer in determining outcome. Since there can be no result without someone to observe it, he concludes, "the physical event and the content of the human mind were inseparable."

What an experiment like Schrödinger's cat means to tell us is that, when it comes to observation, what we look for is what we get. That's a gross oversimplification, of course, as is my reduction of the uncertainty principle to a treatise on the limitations of knowability, but in regard to earthquakes, anyway, I think both are apropos. To listen to a figure like Jim Berkland is to see Schrödinger's thought experiment in action, not as a hypothesis, but as an actual force within the world. Berkland's patterns, not to mention the theory he derives from them, are the ultimate example of perception-dictated phenomena: a case of individual events linked and given focus through the intercession of an observer, who uncovers meaning by bringing his own biases, desires, and preconceptions to the fore. The same is true, in a broader sense, of the experiment at Parkfield, which also identified a pattern where there may or may not be one, and set up a prediction on the basis of what was "found." Yet if all this seems like a conclusive argument for randomness, the beauty of quantum mechanics is that it frames the issue in far more murky terms. Quantum theory doesn't say there are no patterns, just that we cannot definitively recognize them because they are susceptible to an untold range of influences, including our own impulse to observe. The physical world has an order, but it's an order defined by probability, with room for wild cards, subjectivity, happenstance. For all that this sounds like another contradiction, it isn't, really, since once we bring randomness into the larger equation, chance and order are no longer inimical; they become interrelated, balanced, part of the same construction, like the dynamics of yin and yang. As such, uncertainty and chaos theory may represent nothing so much as a passage to reconciliation, by offering a framework from which to take a more expansive view.

Such a view, it occurs to me, is more than a little related to geopoetry, with its emphasis on intuition, on shifts in intention and imagination, on trying to navigate a middle path between a long- and short-range outlook on the world. Perhaps that's why it's so compelling, or maybe it's just one more illusion, I'm not sure. Perhaps in a seismic landscape, this is all we can expect, to look for possibilities in the face of the earth's great intractability, to see how even our most basic observations become inextricably interwoven with all we do not know. Perhaps it even accounts for what I want to conceptualize as the miraculous, for moments like the dream about the World Trade Center falling, for all those strange and edgy confluences we can't explain. Either way, uncertainty might provide a mechanism to push beyond the limitations of both mainstream seismology, with its abiding focus on the practical, and prediction, which too often seems a matter of seeking easy solutions—by allowing us to interpret them not in terms of duality or contradiction, but as smaller pieces of an inherently unknowable whole. What is the shape of this integration? And how does it apply to living in an earthquake zone? The more these answers continue to elude me, the more I think they have to do with the loose ends, the volatility of all those ancient seismic patterns, those stately patterns of geologic time.

One Saturday morning, I find myself in the first car of a Metro Red Line subway, rolling from Hollywood to Universal City, directly underneath the Cahuenga Pass. To my right, Noah presses up against the front window, peering forward down the tunnel as he pretends to be the driver, while behind us sit the usual array of public transportation riders—parents with babies, pensioners, weekend shoppers, tourists, all caught between proximity and distance, between the slow suspension of this moment and the anticipation of arriving further on. To ride the subway in Los Angeles is its own form of dislocation; this is a driving city, after all. Yet despite the contradictions, the experience is oddly comforting, because it brings me face to face with the complexity, the chaos, of this human world. Among the common critiques of Southern California is that it's a segregated society, that people do not mix but stay in their backyards. On the subway, though, you see *everybody*: black, white, Latino, Asian, male and female, young and old. This is one reason I like it down here, the way it reminds me of just how much resides below the surface, of all the overlap, the connections we do not generally get to see. Still, it's something of a tenuous balance, since at a certain point, I always end up remembering that I'm in a concrete tube burrowed under hundreds of feet of shifting rocks and soil and sediment, all of which feels as if it's poised to collapse at the first substantial slip along the fault. I used to think about this a lot back when I lived in San Francisco, used to clasp my hands together every time I boarded a BART train. And today, I feel that same slow prickle of apprehension, a feeling somewhere between uneasiness and anxiety, as I quickly offer up a silent whisper: Please, not here, not now.

The subway hisses to a stop at Lankershim Boulevard, and as the doors slide open, most of the passengers, including Noah and myself, disembark. I gasp a quick sigh of relief to be out of the tunnel, to reenter a world of space and light. Although we're still below street level, the station here is open, airy, with a high vaulted ceiling arcing up above us like a carapace. In the middle of the platform, a public art project—words and images etched on tiles, and installed across a succession of pillars—tells the story of this site, once known as Campo de Cahuenga, where, on January 13, 1847, General Andrés Pico surrendered California to John C. Frémont, effectively ending the Mexican-American War. That's another piece of subterranean history, another layer, a bit of texture, a story most people in Los Angeles don't know any longer, if, in fact, they ever did. No matter where you go in California, you find settings much like this one, places where the past has been eclipsed, where there should be monuments, but you see only streets or subway stations instead. This is why so many people characterize the state in terms of erasure, of forgetting, although to me, the more salient point is that history will always reappear. "Scratch the surface a little and the desert shows through," Bertolt Brecht wrote in 1941 about Southern California, and sixty years later, that still evokes the very essence of the landscape, the idea that here, the past is always lurking, always waiting, covered over but still present, hidden in the least regarded hollows of our lives.

Most days, I would stop and point the project out to Noah, talk about Campo de Cahuenga, trace the fragile filaments of time. This morning, however, Noah and I are in no mood for history; we're Universal Studios bound. For months, I've been promising to take him here, and as we mount the escalator and ride to daylight, he chatters animatedly about what he wants to see. "Is this where they really make the movies?" he asks, and when I say yes, his face splits into a toothy grin. I'm excited, too, but for a different reason—I want to go on the studio tour, because they fake the Big One in a subway station much like this. In some ways, it's like going to another shake test, except that this time, we not only get to witness the earthquake, but also to experience it for ourselves. That, as Noah has pointed out, is what was missing from my visit to San Diego, and if a theme park attraction is more elaborately constructed than a scientific experiment, what does that say about the intersection of myth and reality, the manner in which fantasy and fact collide? Again, a skeptic might just chalk it up to California, but I see the whole thing in more fundamental terms. No, like that "Welcome to L.A." photo, a simulated earthquake is nothing so much as a coping mechanism, an expression of bravado, an attempt to mediate seismicity, to reduce it to a human level, a strategy for making it our own.

Outside, Noah and I cross Lankershim Boulevard and take a tram up to the Universal parking lot. There, we navigate the Saturday crowds. All around us are the signposts of illusion, from the ersatz urbanity of City Walk—an outdoor mall adjacent to

Universal Studios whose prefabricated storefronts, dancing waters, and garish, jutting neon look like a set designer's nightmare of what a city street might be—to the performers who entertain while we wait for tickets, Laurel and Hardy look-alikes in full character regalia, flipping their ties and looking sad around the eyes. Inside, it's both more and less of the same: the first thing we see is an enormous replica of the shark from *Jaws*, hanging upside down, mouth gaping, as if it's just been caught and killed. One more archetypal danger diminished, I think, one more collective terror contained. Noah sticks his head in the shark's mouth, and I take a picture; then, we move on to an enclosed pavilion where kids shoot plastic balls at one another, and from there to an enormous water park, with slides and levels and floor-mounted supersoakers, as well as a huge elevated bucket that spills over periodically, drenching everything within a thirty-foot swath. I stand off to the side, watch Noah run in and out of the water for about twenty minutes, before I wave him over and suggest we take the tour. "What do they have?" he asks, his voice rising, eyes going narrow, slightly skeptical. "What are we going to see?" I tell him the tour will take us all over the studio, that we'll see special effects, rainstorms and mudslides, the smoke and mirrors of this manufactured world. "I don't know," I say, "there might even be an earthquake." And for the first time, I get a little taste of what it's like to be a predictor, hinting at possibilities, offering projections of the future, tapping into someone else's anticipation and desire.

After all, at the mention of the word *earthquake*, Noah gets excited. "A big one?" he asks, and when I shrug my shoulders, his eyes begin to gleam. It's strange: here he is, seven years old, a lifelong Californian, and the only tremors he's gone through have been at the lowest levels of conscious reckoning, like that pair of small September quakes. This is the other side of earthquake country, the way periods of activity are intercut with long interludes of quiet, which lull you into thinking that the planet is somehow stable, that there is such a thing as solid ground. In the seventy-five years after 1906, for instance, just one Bay Area earthquake reached magnitude six or greater, which means that many people lived entire lives beside the fault line without ever coming into direct contact with a seismic threat. I think about this as Noah and I take seats in the open tram and visit the New York streets and the European village, as we cross a mechanically collapsing bridge and watch a flash flood inundate a Mexican mountain town. It's like one last overlay of illusion, a final image of how seismicity eludes us, the idea that, in California of all places, we would have to come to an amusement park in order for Noah to undergo his first significant quake.

The tour continues through the backlot, highlighting all the familiar landmarks of this parallel world. We part the Red Sea and watch a shark attack, see the house from *Psycho* and the cul-de-sac where Beaver Cleaver lived. After half an hour, I start to feel the faint pull of uncertainty, to wonder if I've got it wrong. We've been all over the studio, and no earthquake appears on the horizon; could it have been replaced by a newer installation, something like the mummy's tomb? I look around to check the possibilities, but the street we're on is lined with soundstages, beige warehouse-style buildings that give nothing away. At the front of the tram, the guide begins to discuss continuity, warning us that we're about to enter a live set, and that if we touch anything, it could disturb the actors, or even sabotage the film. Another complex system, I think to myself, another case of randomness and influence. Then, the soundstage doors slide open, and the tram rolls forward, and before I know it, we're pulling up alongside a subway platform, while our guide announces that we've just arrived in San Francisco, at the fictional Waterfront Station, somewhere underneath downtown.

As the tram comes to a stop, I sense a flutter of anticipation, like I'm sitting on the edge of something, something I wasn't sure I'd see. This is it, I think, and I'm relieved to have made it, but once the moment settles, relief slowly gives way to confusion, and I wonder what exactly *this* might be. I'm not the only one; while our guide chatters on about the food at the Craft Services table, Noah, who's ridden BART more than once himself, leans over and informs me that this station doesn't look at all like the real thing. He's right, of course—the ceiling is too low, the tracks too close, too narrow, and there's a run-down quality to the whole construction, as if the walls themselves were saggy, broken, tired. I remember the tension I used to feel in San Francisco, the way that every BART excursion came accompanied not just by prayer, but by a succession of mental images, a private movie of my own most ghastly imaginings: the rumble of the earth and the shaking, always the shaking, as everything caved in. Still, although I try to bring back the sensation, it remains beyond my access, much like the memory itself. This is the problem with illusion; no matter what, you always know that it's not real. I look at Noah, but he's gone back to checking out the station, so I turn my attention to the tour guide, and as I do, the earthquake hits.

When I say earthquake, I mean earthquake, for fake or not, it takes me by surprise. It doesn't matter that I know it's coming; one second, I'm sitting in my plastic tram seat, waiting, and the next, I'm being jolted into Noah, who jumps at the onset of the tremor and gives an involuntary gasp. My sense of being caught off guard, though, is as short-lived as it is unexpected, ending once the shaking starts in earnest, a prolonged and jagged shaking roll. It feels a little bit like Northridge, except the movements are too uniform, too regulated—big and blocky, with none of the organic flow, the variations of pitch, of roll, of intensity, which mark a natural temblor in the middle of its throes. "See?" I say to Noah. "I told you there might be an earthquake." He smiles briefly, a fleeting flash of teeth, but his eyes are darting, nervous, and when the tram starts rocking side to side along the trackbed, he wraps his body around my arm. "It's okay," I whisper as he presses against me. "It's just a special effect. Everything is fine." Up front, our guide tries her best to play the moment, although she overdoes it; "Everybody stay calm," she yells, as if we were in a real disaster. "In California, this happens all the time." As she speaks, her face widens in mock horror, but a telltale grin flickers, mothlike, at the corners of her mouth.

And I don't know if it's the setting, or the quality of the shaking, or if it's some indefinable conflation of the two. But as the earthquake starts to move in earnest, I begin to feel, well, disenchanted—or maybe disassociated is a better word. A simulated temblor may help us exert control over the uncontrollable, to take an open-ended fear and frame it in terms we can understand. In the end, however, it's no more engaging than a shake test, except you get to stand inside. Last year, when I came back from San Diego, I thought this was the issue, that my inability to go through the earthquake, to *feel* it, had made me stand apart. Today, I have the same sensation, the same inescapable edge of distance, as if I'm watching something happening to someone else. Partly, this has to do with Noah, with my desire to keep an eye out, to make sure he's okay. Nonetheless, as the earthquake continues, Noah's nervousness evaporates; he loosens his grip on my arm, and starts to look around. Certainly, there's plenty to observe here: while the tram slams back and forth, the roof splits open, revealing a jagged patch of pavement and a row of storefronts, including a Chinese restaurant. "Hey," Noah says, eyes bright as flashlights, "it's the street above the subway," and as he leans forward, the tarmac tilts and, in one piece, slides into the station, the motion so slow and even it's like a giant lever has been thrown. Once the street has fallen, an oil truck begins to roll inevitably towards us, slipping down the pavement as if on an invisible rail. Every bit of movement seems to have been choreographed, up to the moment the truck collides with a pillar and bursts into flame. We feel a flash of heat, of light, but it recedes when a train appears on the opposite track. Horn blaring, it plows into the truck and derails, rolling across the platform in another controlled skid. A water main ruptures, and the station begins to flood.

By now, Noah is bouncing right along with the earthquake, as if this were a roller-coaster ride. He's not scared, but neither is he awestruck; rather, he seems to take the experience in stride. As a parent, this is what you hope for, but even so, I can't help feeling that he's disappointed, that this temblor hasn't done what it was meant to do. Initially, I think, that's because of the flimsiness of the illusion, the way that, for all its vivid iconography—the collapsing street, the exploding tanker, the subway derailing, those elemental images of devastation—this event is almost entirely smaller than life. It's not just the mechanical pitch and yaw of the shaking, but the insubstantiality of the effects, the way the burning truck looks like a three-dimensional image on a billboard, while the train is an aluminum tube on shopping cart wheels. The real reason, though, becomes clear just before we leave the soundstage, after the earthquake is done. We're in the tram, waiting—for some kind of clearance, word that they're ready at the next stop—when the subway station starts to come together, as if time itself had magically reversed. First, the train pulls back onto the track and withdraws from the station; then, the tanker returns to the upper level, and in one smooth motion, the ceiling closes off the street from view. Gone is the flooding, gone the fire and destruction, gone every bit of evidence that anything has happened here at all. It's the perfect caricature of a miracle, in which, at the touch of a finger, the past becomes retrievable, and events rewind with the regularity of a clock.

And just like that, I understand what's happening, why I wanted to see this earthquake, and why I'm leaving unfulfilled. What we've just experienced is the ultimate mechanical earthquake, the earthquake the predictors mean to claim. It's an event devoid of randomness, one that repeats at well-defined intervals, in patterns that are utterly deterministic, so controlled that, when it's finished, it actually erases itself. To some extent, this is a metaphor for how we live in California, how we shrug off seismicity, turn it into rides and stories, reconstruct collapsed buildings and fallen freeways and try to forget the way they looked before. But if that's the case, then it's a metaphor with no depth, no history, a metaphor of a never-ending present tense. This, you could argue, is the point entirely, the very essence of a mediated earthquake, that it is something we can box up and name. You can see it, you can quantify it; it has no shadows, no unexplained correlations—everything exists in black and white. Still, while it's interesting, even occasionally surprising, to sit in an imitation subway station and watch it rumble, in the end, it tells us nothing of ourselves. There is no danger, no sense of chaos: in short, no question about how the earthquake ends.

If I ever doubted that we need this, I'm given one more reminder towards the end of the day. Noah and I are back in the subway—the *real* subway—standing on the Lankershim platform, waiting for the Red Line to Hollywood. Ahead of us, the tunnels yawn black and tight, pathways cutting through the hills. The station, meanwhile, seems to float like a cathedral, an enormous empty space in which I pray. It looks nothing like the set at Universal, but that doesn't mean it's more stable; in fact, it feels both more and less substantial at the same time. So once again, I clasp my hands and whisper: Please, not here, not now.

Noah watches me for a minute, then draws a deep breath. "Dad?" he says, and when I nod, he asks me, "What would happen if there was an earthquake down here?" I take a good long look to try and gauge him, although his eyes are hard to read. But finally, I have nothing but uncertainty to offer. . . .

"I don't know," I tell him. "I don't know."

SHAKING ALL OVER

The North Anatolian Fault extends for nearly a thousand kilometers across the northern edge of Turkey, in a line just south of the Black Sea. Running on an east-west axis, from the vicinity of Erzincan to the Sea of Marmara, it is what seismologists call a "master fault," the defining geological feature of its region, visible in many places at the surface, like a long scar cut into the earth. Along much of its length, the fault passes through highly populated and industrialized areas, which means that it poses a considerable risk. To understand exactly how considerable, we need look no further than the North Anatolian's two most recent ruptures, which took place in the summer and fall of 1999. On August 17 of that year, a 7.4

earthquake devastated Izmit, about fifty miles southeast of Istanbul; less than three months later, on November 12, a second temblor, measuring 7.1, struck the city of Düzce, seventy miles further east. In their 2000 paper on the two disasters, USGS seismologists Tom Parsons, Ross Stein, and James Dieterich, along with Aykut Barka of Istanbul Technical University and Shinji Toda of the University of Tokyo, estimated that the earthquakes "killed 18,000 people, destroyed 15,400 buildings, and caused $10–25 billion in damage."

Of course, the trouble with such figures is that they blur to indistinction, becoming abstractions rather than representations of actual damages, actual lives. This is a common problem when it comes to earthquakes, which, by our acts of calculation, lose a good deal of their elemental force. How do we wrap our minds around eighteen thousand deaths, fifteen thousand buildings? How do we make those numbers real? Eighteen thousand people is a town, or at the very least a neighborhood, a whole community shaken clear of the earth. For me, there's only one way even to try and comprehend it, which is to think in terms of specifics, particular details or images that give the devastation shape. In the aftermath of Northridge, I focused not on statistics—body counts, dollar totals—but on the moments I was part of, the situations I had seen. I remember my neighbors pouring out to the street in nightgowns, T-shirts, bathrobes, turning on their cars to listen to the radio, chattering in nervous clusters like so many disoriented birds. With Turkey, I engaged in a similar process, albeit from a greater distance, as if this might give the quakes there a dimension I could understand. Rather than try to make sense of them through facts and figures, I conceptualized the disasters in terms of a single image, an aerial photograph that appeared the morning after the first tremor on the front page of the *New York Times*. In this picture, an apartment house in Istanbul lies half-collapsed beside a broad boulevard, debris spilling into the intersection like blood flowing from a wound. It's a modern structure, five, maybe six stories in the sections that remain standing, unadorned except for an eruption of little balconies, and it looks eerily like a building I often drive past on the corner of Los Feliz and Griffith Park boulevards. So this is what will happen, I recall thinking, this is what the future holds. Even now, I can't escape that pulse of recognition, as if what I'd seen were less a photo than a premonition of the devastation that's to come.

As it turns out, my conflation of Turkey with Southern California is, seismologically speaking, more than just a matter of imaginative association, although I came to understand this only after the fact. The North Anatolian Fault, after all, is regarded in geologic circles as a sister fault to the San Andreas, a kind of mirror image on the other side of the world. The San Andreas may run north-south, not east-west, but that's a negligible distinction. More to the point, it and the North Anatolian are the same length, and, as faults go, relatively linear; they also slip at an equivalent long-term rate. The North Anatolian even has a creeping section in the middle, much like the San Andreas. Perhaps the most essential correlation, though, has to do with Los Angeles and Istanbul, both of which butt up against the seismic firing line. Istanbul may be a little closer—barely ten miles from the North Anatolian at its nearest point, as compared with the fifty miles separating L.A. and the San Andreas—but the cities share some striking physical characteristics, most notably the way each sits atop a sedimentary basin, and is therefore subject to the amplification of seismic waves. When you add the fact that Istanbul is almost exactly the same distance from Izmit as Los Angeles is from San Bernardino, where many geologists expect the next great southern San Andreas earthquake to begin, the parallels become too compelling to ignore. It's hardly surprising that within twenty-four hours of the August 1999 temblor, response teams from the USGS and the Southern California Earthquake Center were on their way to Turkey to investigate. As UCLA geophysicist David Jackson told the *Los Angeles Times*, "This is a special earthquake. Big ones are rare, and we need to understand the big ones, so this is very, very important. It is in many ways like Los Angeles."

Still, for all the similarities between the North Anatolian and the San Andreas, there is one surprising difference: in the last hundred years, while the San Andreas has yielded two earthquakes of magnitude 6.7 or higher, the North Anatolian has generated twelve. More astonishing, in the geologic nanosecond since December 26, 1939, when a 7.9 killed thirty thousand people in Erzincan, there have been thirteen 6-plus quakes on the North Anatolian system, eight of which comprise what appears to be a sequential east-west pattern encompassing virtually the entire length of the fault. From Erzincan, approximately four hundred miles east of Istanbul, the progression extends first to Erbaa, which suffered a 6.9 on December 20, 1942, and Tosya, where a 7.7 hit the following November 26. On February 1, 1944, a 7.5 shook Bolu-Cerede, although after that, it's thirteen years until the Abant quake of May 26, 1957, which measured 6.8. Then, on July 22, 1967, a 7.0 in the Mudurnu Valley, and finally, the two temblors of 1999. If you look at a map on which all these earthquakes have been marked off, it's almost like seeing the results of a chain reaction, or what Ross Stein calls "a domino effect." Each event picks up precisely where the previous one left off, until you have an unbroken line stretching from Erzincan to Izmit and beyond. How do we account for this? And if the faults are comparable, why does the San Andreas not behave in such a way? In a 1997 paper coauthored with Aykut Barka and James Dieterich, Stein proposes the following explanation:

> We suggest that the propensity of the fault zone toward progressive failure is a product of its simple, straight geometry, which . . . makes for efficient transfer of stress; its isolation from other faults, which minimizes stress transferred between the North Anatolian and competing faults; and its . . . echelon character, which tends to keep the entire fault from rupturing at once. By contrast, the San Andreas fault, which lacks an historical record of progressive shocks, produces larger earthquakes along its smoother trace, and generally lies close to other major faults, making the stress transfer more irregular and complex.

On the one hand, such a statement seems nothing if not contradictory, tracing what look like fundamental variations in the two faults. How can we possibly call them similar when the findings of Stein and Barka and Dieterich suggest they couldn't be less alike? The more I think about it, however, the more I see it as yet another issue of balance, another instance in which the seismic landscape defies dualistic thinking, another example of the uncertainty, the fluidity, of this ever-shifting world. Yes, Stein and his colleagues mean to tell us, the San Andreas and the North Anatolian are analogous, but they are also very different, and there is no real conflict between those ideas. Instead, we need to focus on the larger picture, the broader patterns, the way that outside factors—the terrain, the existence of surrounding faults—can influence what happens in the ground. Seismicity, in other words, is a complex system, and if we want to come to terms with it, we have to take a wider view. To a large extent, what we're dealing with here is chaos, although as in theoretical physics, it's a chaos that reflects a certain order, even if the order is one we cannot see. No fault exists in a vacuum, and once we grasp that, perhaps it can lead us to a whole new framework for thinking about earthquakes, and the subtle ways they intersect.

Ross Stein occupies an office just down the hall from Allan Lindh at the USGS's Menlo Park field office, but it might as well exist in a different universe. Unlike Lindh's elaborately disordered work space, Stein's is controlled, uncluttered, with well-organized shelves and file folders, and a desk that's mostly clear except for family photos and a sleek black Apple laptop on which he runs computer simulations to support his points. Stein himself is pin-neat also, clean-shaven, short gray-black hair evenly parted, dressed in dark slacks and a madras button-down, sleeves rolled crisply to the middle of his arms. Late on a Monday afternoon, he sits in a swivel chair, sipping a cup of tea, talking as the autumn sun grows thin outside his window and the day evaporates beneath the creeping shadows of twilight. Even his voice marks him as fastidious—slow, measured, every word clearly enunciated—as do the occasional pauses with which he punctuates his conversation, as if to make sure he is understood.

Despite his air of quiet consideration, Stein is, in his own way, as much of an iconoclast as Lindh ever was—maybe even more. Since the early 1990s, he has spent the bulk of his time immersed in the nebulous territory of earthquake interaction, an area of inquiry so new (on both the human and the geologic level) that it has yet to be codified in any but the most general terms. At the basic level, interaction appears a simple process, but as always when it comes to earthquakes, appearances can be deceptive things. The idea is that, when a tremor takes place, it not only relieves stress in the area that has broken, but also transfers it to adjacent sections of the fault. Those segments, in turn, are brought closer to their own potential failure point. This seems to be what's happening on the North Anatolian, to cite an obvious example, but although the pattern there goes back to the 1930s—further if you factor in apparent progressions in the eleventh and seventeenth centuries—interaction only emerged as scientifically credible in the wake of the Landers earthquake, which broke at 4:58 a.m. on June 28, 1992, along a previously undiscovered fault in Southern California's Yucca Valley. Although at 7.3, it was (and remains) the strongest California temblor in the last half century, Landers is a relatively little known earthquake; occurring in a small desert community, causing only one fatality, it doesn't linger in our consciousness as Northridge or Loma Prieta do. Still, in many ways, Landers may be the most significant of the three tremors—if only for what happened afterwards. Barely three hours later, a 6.6 shook the mountain resort of Big Bear, sixteen miles northwest of the initial shock. This quake, too, struck on an uncharted fault that was eventually found to run into both the Landers section and the San Andreas, forming a seismic triangle ten kilometers below the surface of the earth. "Even more remarkable," notes Philip L. Fradkin in *Magnitude 8*, "the Landers quake set off seismic activity as far away as Yellowstone National Park, 750 miles to the northeast. Faults thought to be separate were now determined to be interconnected."

It's one thing, of course, to say that faults are interconnected, and another altogether to explain what this might mean. Here again, we find ourselves pressed up against the limitations of seismic history, especially in California, where the documentary record goes back only so far. For Stein, whose earliest interaction efforts had to do with Landers, that's why the North Anatolian is so important, although, he admits, he had no particular sense of this before he started to study the Turkish system in 1995. Instead, his investigations there, "like almost all work, had a kind of accidental beginning"—when, at a scientific conference, he met Aykut Barka, who told him, "You're doing great work, but on the wrong fault." A few years later, after the collapse of a European field trip, Stein called Barka, and they embarked on what has become nearly a decade of collaborative research. "You know," Stein says, smiling at the elliptical dance of circumstance that has led him from the Southern California desert to the Bosporus, "we're scavengers. We go wherever there's fresh carrion. And Turkey . . . well, twelve large earthquakes in this century means, inevitably, that it's a better laboratory for studying large earthquakes. Large earthquakes are what cause damage and what kill people, and that's really the business we're in. What matters to us are the very large earthquakes which, on both faults, recur every several hundred years. So if you're going to sit down and try to understand a pattern of behavior, you're not going anywhere with a 200- or, in the case of California, 150-year history. In Turkey, we can go back 1,500 years."

When Stein refers to a fifteen-hundred-year chronology, what he's talking about, primarily, is Istanbul, which has suffered severe seismic damage at least a dozen times since the middle of the fifth century, most notably on September 14, 1509, when upwards of thirteen thousand people died in an earthquake estimated to have measured 7.0. Elsewhere along the North Anatolian, information is just too sparse, too sketchy; it's impossible to determine whether there are no reports for a given temblor because no one in the area felt it, or because, as Stein suggests, "an entire city was destroyed and nobody made it out alive." Even in Istanbul—as is the case with any ancient culture—the evidence is shaky, a loose amalgamation of eyewitness accounts, damage reports, and sheer invention that a present-day interpreter can't help but take with a substantial grain of salt. If this isn't quite the realm of, say, New Madrid, it's pretty close, which means that, like Susan Hough, Stein must read between the lines. "A lot of reports," he explains, "are exaggerated because kingdoms and individuals hoped that if they exaggerated the damage they'd get more money to repair it. Or there are people who visited the Ottoman Empire and went back to the doge of Venice or a Saxon king, to report to their principality what they saw. It's the worst kind of data you could ever want to make an earthquake out of." Nonetheless, the least accurate account can provide a starting point, a window, however opaque, onto the past. To highlight this, Stein brings up on his computer a picture of Istanbul drawn after the 1509 earthquake, in the corner of which stands the dome of Aya Sofia, built in the sixth century, and once the largest church in the world. "This," he says, "illustrates the yin and the yang of what you've got, because it's a pretty good record of damage to walls and towers, some of which are still there." He points to Aya Sofia, to what appear to be cracks, and damage to the dome. "However," he goes on, "there's also a meteorite flying by, and that has to be an exaggeration. This earthquake didn't trigger any meteorites. So that calls the whole thing into question. That's what you have to deal with. That's what you have to sort through."

In many ways, Stein's sense of yin and yang, of sorting through, suggests another strategy for reading anecdotal information and using it to expand the range of seismic history. It's also something of a metaphor for the whole process of earthquake investigation, in which even the most rigorously gathered data is subject to interpretation, and everyone has a different opinion about what things mean. Contemporary seismology, after all, may be less anecdotal than a sixteenth-century drawing, but that doesn't mean it's necessarily any more clear. This is especially true of earthquake interaction, which, since it's still so raw, so undeveloped, is the subject of intense debate. One criticism of the theory is that the stress changes on which it depends are too small to trigger earthquakes, although advocates argue that when a fault is close to failure, it takes very little to push it past the point of breaking—"perhaps nothing more," Susan Hough has written, "than the final grain of sand landing atop the increasingly unstable sandpile." Yet even if you accept Hough's analogy, there's still the question of how interaction works. With Landers, the process appears (at least nominally) immediate, triggering earthquakes in a matter of hours. On the North Anatolian, however, we're dealing with long-term connections, which can seemingly remain dormant for decades. This, too, has become an area of contention, for interaction shouldn't function like that. "If I ratchet up the stress here," Stein explains, pointing to a seismic map, "and you're the next fault down the line, the odds are you're going to rupture soon. I'm still able to trigger you years from now, but the likelihood slowly diminishes over time." As a result, some geologists dismiss the North Anatolian as a freak, an anomaly—a classic falling domino sequence, yes, but one that isn't replicated anywhere on earth. Can we frame a general rule, these critics ask, from such an idiosyncratic circumstance? Or are interaction theorists simply "shopping around" for earthquakes to support their point of view? "This," Stein says, "is a legitimate criticism, because a lot of science is done on a case by case basis. You go where the data is. But in my view, the fact that it occurs at all is exactly why we should be studying it. We have falling dominoes. Why do they fall?"

It is here that Turkey's fifteen-hundred-year earthquake catalogue becomes essential, for when we examine it closely, several issues assert themselves. Most important are the two prior cases of seismic triggering, the first involving large earthquakes in 967, 1035, and 1050, and the second the sequential rupture of about four hundred miles of the fault between 1650 and 1668. What makes these apparent interactions compelling is that, unlike the recent east-west pattern on the North Anatolian, they run west to east. That may seem like another dichotomy, but if we widen our focus, the situation takes on unexpected confluence. After all, of the earthquakes to strike the North Anatolian in the last sixty years, four—in 1949, 1966, 1971, and 1992—have broken in an L-shape east of Erzincan, creating a counterprogression to the main movement on the fault. This suggests that, on the North Anatolian anyway, interaction operates in both directions, and as to what *that* means . . . Stein believes, it could be the key to everything. "Here's what I wanted to show you," he says, voice rising in excitement as he taps a silent code of symbols into his laptop. After a moment, a PowerPoint map of the North Anatolian unscrolls across the screen. "Okay," he continues. "Here's the fault, and on top of it, we're going to put each earthquake as it occurs, and its length will show up here, and the amount of slip, the amount of ground displacement will be plotted here. All things being equal, the places where stress increases are going to be the sites of the next earthquakes. As stresses accumulate, I want you to look at the ends of the fault. Once an earthquake occurs, the stress relaxes in the middle. But beyond both ends, we've actually brought the region closer to failure. Because the fault is a simple straight line, it jacks up the stress at either end."

Stein presses more buttons on the keyboard, and the map on his computer slowly stutters to life. While I watch, as if from a celestial distance, the North Anatolian breaks in a stately progression of images, temblors blooming like bouquets of electronic flowers. In fact, bouquets are exactly what they look like, for as the quakes follow one after the other, Stein's software highlights what are known as stress fields, represented here by vivid petals of color sprouting out in all directions from the fault. In areas where stress decreases after an earthquake—along the lengthy sides of the fault, mostly—the computer shows large fronds of vibrant blue or purple; where it builds, meanwhile, I see spears of electric red. First 1939, then 1942 and 1943 and 1944, and on up through the years to Izmit, while in the other direction, a second series of tremors breaks in 1949, 1966, and 1971. Sure enough, with each earthquake, the system does just what Stein claims for it, stress fields going blue as strain diminishes up and down the flanks of ruptures, while literally bursting red at both the eastern and western ends of every slip. "See?" he says, once we've clicked through the entire sequence, his tone by turns elated and content. "The two ends have, as far as we can tell, an identical likelihood of rupturing." He taps another key, resetting the program. The blossoms of red and blue disappear as if never there.

The simulation I've just witnessed offers a striking visual breakdown of the dynamics of interaction, the way energy is transferred from segment to segment along a fault. You can see it, see how the pressure rises and falls, following its own circadian rhythm, like the beating of a seismic heart. At the same time, all this raises one more question, which is why, if stress mounts equally at the two ends of a rupture, interaction mostly travels in one direction or the other, rather than both at once. Even in the wake of the 1939 Erzincan earthquake, which appears to have triggered at least some slippage to either side, the westward sequence is so much more prominent that it reduces the smaller easterly progression to an afterthought. How does this happen? Is it simply a matter of momentum? Here again, no one knows. The best we can do is to theorize some kind of clamping mechanism that impedes movement on one side of the fault. "I don't know why it's happened," Stein says, "but as long as it has, the stress on the unclamped side tends to promote the next rupture." He fingers his keyboard again, sets up another progression. With every click, the North Anatolian breaks anew. "You can see," he tells me, identifying each eruption of color, "that if you squeeze the fault at one spot, it gets things going in one direction." I nod, and stare as earthquakes tumble like falling stones.

For the next few minutes, Stein directs the simulation, stopping on certain frames to highlight different bits of interaction, while noting the accumulation of stresses that marks the hidden history of the fault. At one point, he runs a demonstration that focuses in on Istanbul and Izmit, then sets the program clock to 1900, so we can observe the buildup to the quakes of 1999. "Look how Izmit slowly turns redder and redder as the twentieth century progresses," he enthuses, "until this point, when we have the earthquake." He resets the software once more, then a third time, his touch upon the keyboard like the hand of God. Sitting here, I'm reminded of Edward Lorenz, whose discovery of the butterfly effect grew out of a similar experience, running and rerunning simulations to see what patterns might emerge. According to James Gleick, Lorenz first noticed anomalies in his data after leaving his office for a cup of coffee; on his return, an hour later, a whole new sequence had been spawned. Briefly, I consider what might happen were Stein and I to do the same thing right now, to step outside for tea or coffee. What would we see when we got back? Stein, however, has a mug of tea already, and anyway, we're not looking at an open system, in the manner of Lorenz's weather models, but a PowerPoint presentation, an electronic rendering of the past. There are no possible projections here, no speculation, only the inexorable movements of time gone by. In a certain sense, that makes Stein's simulation the opposite of Lorenz's, since, stopping as it does after the Düzce earthquake, this is a simulation where the future never quite arrives.

Still, watching Stein put the North Anatolian through its paces, I begin to feel a deep and unanticipated wonder, like wings unfolding in my chest. It's the scope that does it, the idea that here, within a single minute, I can witness a hundred, five hundred, fifteen hundred years of earthquakes ripple back and forth across the screen. The sensation is oddly similar to climbing down into the San Andreas, a visceral apprehension of eternity, of the depths of geologic time. After all, if a demonstration such as this has anything to offer, it's a structure for making history accessible, a way of enlarging our perspective, of allowing us to frame seismicity in the broadest terms. To see a fault rupture repeatedly over a span of centuries (even if they're virtual centuries) is to transcend the limits of chronology, to confront not just an earthquake but the whole concept of *earthquakes,* to recognize a continuum, and the larger order it implies. "Once you give up on the idea that an earthquake is an island," Stein says, "you can see earthquakes as a sequence, as a process. And you begin to be attuned to how they turn each other off and on." As if to prove the point, he takes me through the demonstration one last time, and again, I note the breaks, the patterns of stress and movement, transfiguring the fault line with a logic all their own.

The question, of course, is the extent to which that logic can be applied to any other context. Is the North Anatolian, in other words, really an exception, or is it just a place where interaction is more pronounced? This is a defining issue, and in order to resolve it, we must go back to the complex web of schisms that comprise the San Andreas Fault. Although to see the San Andreas on a map is to imagine it as one long, sweeping curve that, for the most part, mirrors the coast of California, in truth it's less a single line of demarcation than, in the phrase of *California Fault* author Thurston Clarke, "a broad, complex zone of parallel cracks, ranging in width from a few hundred feet to almost a mile." Further complicating matters are the adjacent "major faults" Stein, Barka, and Dieterich have cited, from the Hayward, Calaveras, and San Gregorio in the north to the latticework of scarps—including the Newport-Inglewood, the Hollywood, the Santa Monica, the Sierra Madre–Cucamonga, and the San Jacinto—that underlie Southern California's rock-slid slopes and crumbly terrain. Los Angeles, in fact, is unmatched among our cities in regard to the number of faults running beneath it, and if interaction tells us anything, it's that they all exert an influence on one another, keeping the entire area in a state of constant tension, a delicate balance that can be tipped by the slightest pressure, the slightest movement on any fault. For an example of how this works, we need only look to Landers, which even as it relieved stress on the Calico and Lenwood faults (as well as parts of the San Andreas), increased strain on the San Andreas near San Bernardino. As Kerry Sieh and Simon LeVay point out in *The Earth in Turmoil*, "The next rupture of this segment—a rupture that is expected to produce a great, 7.5–8.0 earthquake—has been brought 14 years closer by the Landers earthquake."

What makes such a statement compelling is the way it highlights the idea of seismicity as a regional phenomenon, in which interaction unspools from fault to fault like an intricately woven chain. That, Stein says, is an essential development, a strategy for bringing the lessons of the North Anatolian to bear on the San Andreas, where seismicity is a more enigmatic process, or at least less well defined. Here again, we need to take the long view, to look for similarities, points of overlap; although we might not see interaction "writ large" in California in the form of sequential breaks along a single fault line, that doesn't mean it isn't there. In fact, Stein argues, "there are a lot of little dominoes" in the system, going back to the earliest recorded temblors on the

San Andreas, from the pair of 1857 Parkfield shocks that preceded Fort Tejon to the fourteen or more tremors measuring 6.0 or higher that rattled the San Francisco Bay Area in the seventy-five-year period leading up to the 1906 quake.

It is this clutch of Bay Area earthquakes, actually, that appears to offer the strongest historical evidence for interaction in connection to the San Andreas, although as is true of Landers, the process is elliptical, moving in anything but a straight line. Beginning with June 10, 1836, when a 6.5 struck near Monterey Bay, you can see a steady string of tremors, ranging from magnitude 6.3 to 6.8, trace a loop around the San Francisco Peninsula like a target or a noose. In June 1838, barely two years after the Monterey Bay jolt, a 6.8 shook the San Andreas south of San Francisco; it was followed by the two Great San Francisco Earthquakes of the 1860s, the first of which, a 6.5, broke only a few miles north of Loma Prieta on October 8, 1865. So closely does its epicenter match the 1989 earthquake that some people consider them related, a pair of mirror shocks reflecting back upon each other across a distance of 124 years. This event, along with the 1838 temblor and a 6.3 on April 24, 1890, led Allan Lindh to compare nineteenth- and twentieth-century seismicity on the San Andreas, and to issue his initial forecast for the Santa Cruz Mountains segment of the fault. That's an important moment, yet despite its lingering resonance, I can't help thinking that it fills in only part of the picture, since if the interactive model here is accurate, the San Andreas may be best understood as part of a more extensive network, in which the entire fault structure of the Bay Area is involved.

The trouble with looking at interaction through such a wide-angle filter is that it requires us to undertake a difficult conceptual jump. While, on the surface, it makes sense that earthquakes within a single geographic region should slip in conjunction with one another, the Bay Area is tricky because, unlike Landers, the faults don't all link up. Yes, the Calaveras and San Gregorio cut close to the San Andreas, but the Hayward carves its own path through the East Bay, and other, smaller offshoots, like the Concord and the Green Valley, seem to exist in seismic limbo, isolated segments running parallel to their larger counterparts, individual and distinct. How, then, does a quake on one fault influence activity along another that may stand independent, fifty or a hundred miles away? One of the unsettled questions of Bay Area seismic history is whether the 1865 and 1868 earthquakes are related, even though one broke on the San Andreas and the other on the Hayward Fault. Proximity—both temporal and geographic—suggests some connection, but proximity is not enough. Rather, we need a link, a reciprocal passage, an explanation for how these earthquakes might, as Ross Stein puts it, "ricochet around the bay." Is it possible to transfer stress from fault to fault without a direct route of access, or is this where interaction starts to fray?

The solution, Stein thinks, has to do with what he calls "side lobes"—the bursts of color that, in his computer simulations, radiate outward from the sides of broken faults. On the North Anatolian, they are almost always blue or purple, indicating a substantial dissipation of stress. In the Bay Area, however, we see another kind of interaction, which becomes apparent as soon as Stein brings up a map of the region on his laptop, and sets it back to 1838. "This is the 1838 earthquake," he says, pointing to a spot on the San Andreas, "and this is the Hayward Fault here, parallel to it. The interesting thing is that this 1838 event produces lobes off the end, which may have triggered the 1865 earthquake, but it also produces side lobes, so it may have triggered 1868, as well." Then, he starts up the progression, and I watch it unshuffle across the monitor, another silent glimpse of history, of time telescoped and, perhaps, redeemed. Sure enough, the 1838 shock sends tapers of red exploding north and south along the San Andreas, but it also yields large red side lobes that stretch across much of the Bay Area, stressing up the site of the 1868 rupture, not to mention the likely epicenters of several other quakes. As with any historical recreation, the situation is muddied by the vagaries of nonscientific data; no one knows, for instance, where exactly any of these earthquakes happened, nor how big they were. Still, if Stein's model is not definitive, it does at least seem possible, which implies that interaction could be as wide-ranging as he believes. "Yes," he argues, "there are limitations. But while this is a messier picture, it's plausible that part of the story of the Bay Area is really much more similar to Turkey than we might have thought."

Adding credence to Stein's theory is something Allan Lindh, among others, observed in the early 1980s—the apparent shutdown of virtually all of Northern California in the aftermath of the 1906 San Francisco quake. Between 1906 and 1984, only one six-plus tremor occurred in the entire region, a 6.2 that hit the Calaveras Fault on July 1, 1911. This means that for nearly seventy-five years, Northern California was essentially inactive, or as inactive as an earthquake zone can be. In and of itself, that's not surprising; as early as 1910, Harry Fielding Reid's elastic rebound theory, with its notion of seismicity as a cyclical process, indicated that there would be periods of dormancy after major temblors, as the stress worked its way back to the failure point. Yet if Reid's theory explains the stillness on the San Andreas—the 1906 earthquake relieved centuries of tension, breaking, as it did, over a length of nearly two hundred miles—it doesn't account for all the other faults that went into suspension also, almost as if a plug had been pulled. Rather, such a situation seems to point to interaction, which may tell us not just why certain earthquakes happen, but also why they don't. "What we know," declares Christopher Scholz, who in 1982 cowrote the first scientific paper on the subject, "is that 1906 dropped the stress on both the San Andreas and surrounding faults. The question we need to ask is why. And the answer appears to be that, even as it relaxed stress on the San Andreas, the earthquake generated a field of reduced stress across the region, which lowered the probability of earthquakes on all the parallel faults. So basically, the 1906 earthquake shut those guys down. The stress drop, of course, is lowest right at the San Andreas. It drops less on the Hayward, and less on the Calaveras and these other faults. But for a period of time, it stopped them. It just shut them down."

The process to which Scholz is referring has come to be known as stress shadowing, because of the "shadows" cast by big tremors over the regions where they take place. When I ask Stein about it, he runs one final demonstration, beginning in 1838 and ending just after the 1906 quake. Again, I watch seismicity increase throughout the nineteenth century, one fault triggering another, until the map on Stein's computer has turned a vibrant shade of red. Then, we get to 1906, and the earthquake happens, after which enormous blue side lobes cover San Francisco Bay. Except for two thin tongues of red—or, as Stein calls them, "blow torches"—near Cape Mendocino and San Juan Bautista (the end points of the rupture), there is nothing except blue to be seen. It's as if, on the other side of this earthquake, Northern California has been literally cleansed of all its stresses, returning to a kind of zero-sum condition, a state of original innocence, in which the seismic clock is not merely slowed down, but entirely reset.

Of course, the difficult thing about earthquakes is that there is no zero-sum condition, for the stresses are always changing, building and colliding, pushing one another in an endless tug-of-war. As for the moment when things will pick up, Stein believes Loma Prieta may be an indication that the process has already begun. The same appears true of Southern California, where the stress shadow cast by Fort Tejon has only lately begun to lift. From 1857 until the late 1980s, the Southland remained relatively quiet, experiencing only two significant shocks—the 6.3 Long Beach earthquake, which killed 120 people on March 10, 1933, and the 1971 San Fernando tremor—neither of which involved the San Andreas Fault. Since 1987, however, seismicity has increased almost geometrically, in a pattern Christopher Scholz calls eerily similar to what took place throughout the Bay Area during the decades prior to 1906. Judging from the seismic record, it's hard to argue with Scholz's position: in the little more than six years between the October 1, 1987, Whittier Narrows earthquake and Northridge, Southern California was rocked by six temblors of magnitude 5.9 or higher, although activity has slowed since then. Here, however, we again find ourselves on the precipice of pure conjecture, since, for all that interaction has to offer, it cannot tell us when the next event is coming, nor reveal the fullness of the sequence that lies hidden in the ground. In many ways, this is the issue Allan Lindh confronted in his efforts to understand the Lake Elsman earthquakes; it was not until after Loma Prieta that he could say with any certainty what they might mean. Yes, as Lindh suspected, there was a connection, but the details of it became apparent only in retrospect.

Stein laughs when I bring this up, acknowledging that it's the nature of the business. Still, he continues, the same could be said of aftershocks, which are, at the most basic level, triggered earthquakes, although within a far more narrowly defined range. It's a fascinating comment, for aftershocks have always seemed to me the ultimate refutation of the notion that seismicity is random, arising, as they do, in direct connection to earlier earthquakes, and operating in accordance with specific rules. Even their magnitudes are generally proportional, following the Gutenberg-Richter relation, a scale developed in 1958 at Caltech by which, note Susan Hough and Lucy Jones in a 1997 American Geophysical Union article, "a magnitude 6 mainshock is . . . expected to produce, on average, one magnitude 5 aftershock, and approximately 10 events near magnitude 4," and so on down the line. Of course, as is the norm when it comes to earthquakes, this apparent order may be meaningless. But considered through the filter of interaction, it suggests another layer to the relationships between earthquakes, whether they extend for days or months or even hundreds of years. "In some sense," Stein says, "we're redefining the idea of aftershocks. We're saying aftershocks are just earthquakes that occur where stress increases as the result of a mainshock." He peers at me as if to make sure that I follow, then switches off the laptop and puts it away.

On the other side of the window, the sky has gone dark, a deep blue laced with narrow stripes of purple, and although I strain, I can't quite make out the bare-ribbed trees as they set to keening, rustling in the swirling autumn breeze. Inside, however, I see nothing but connections, tremors begetting other tremors, as if extending in a direct progression from the dawn of time. Again, I feel those butterflies of wonder; again, I ask myself how far all this might go. Is it possible, for instance, to frame Loma Prieta as a long-delayed aftershock of the 1906 earthquake? Or perhaps a distant foreshock to some not-yet-realized event? On the face of it, these are absurd questions, at least in terms of how we currently understand them, but if Stein's work highlights anything, it's that the dynamic among earthquakes may be broader than we think. In many ways, it's a new approach to the same old conundrum, the one about what seismicity does or does not mean. What are these faults trying to tell us? How much can we possibly hope to understand? I still don't know the answers, or even if there are any answers, but as the night continues its slow fade over Northern California, I stare into the blackening distance and feel as if I'm getting close to something, a frame of reference or at least a context, some kind of logic I can almost see.

Loma Prieta is an illogical earthquake. The more I circle back around it, the more I understand this, which may be why I cannot leave it alone. It is the seismic equivalent of an onion, an earthquake that continues to surprise us every time we strip another layer away. Although it struck the San Andreas on a segment that the Working Group on California Earthquake Probabilities had identified, eighteen months beforehand, as the most likely spot for a significant Bay Area temblor, it did not behave as seismologists had anticipated, nor, for that matter, like any rupture in the recorded history of the fault. In the first place, the epicenter did not fall on the fault line, but nearly four miles to the west. Equally confusing, the quake failed to produce the kind of orderly, north-south fissures normally associated with the San Andreas; instead, notes J. E. Ferrell in a 1990 *Los Angeles Times Magazine* story, "the hundreds of cracks [investigators] found—some large enough for a person to stand in, some only 1/8-inch wide—ran perpendicular to the fault or broke in a direction opposite of what was expected. It just didn't make sense." Finally, the tremor generated a highly anomalous aftershock sequence, which, rather than echoing up and down the San Andreas, "spread in a cloud across two minor faults nearby, the Sargent and the Zayante." All this suggests that the architecture of the fault is more intricate than geologists had previously realized. As Ferrell explains:

> The Loma Prieta quake was both a strike-slip (horizontal) rupture and a thrust (vertical) rupture. Starting 11.5 miles underground and extending to within 3.7 miles of the earth's surface, an angled segment of the San Andreas Fault broke as the Pacific Plate moved northwest in relation to the North American Plate *and* thrust up and over it. The angled, thrust motion, never before associated with the San Andreas, explains why the epicenter was miles off the San Andreas Fault line. At the surface, the ground was pushed up 14 inches and north about 7 inches. At the rupture point or hypocenter, the earth moved up about 4 feet and north about 6 feet.

On the one hand, the saga of Loma Prieta tells us that earthquakes will always confound our expectations, no matter what we think we know. In that sense, it's another cautionary tale, an indication of how fast things change in the seismic trenches, a reminder of the half-life of ideas. At the same time, there's a good deal more at work here, for the quake also seems to fit a lot of patterns, both immediate and long-term. I'm not talking now only about the forecasts, nor Stein's sequences of interaction, although clearly, both are part of the larger story, strategies for reading the quake as a chapter in a narrative that extends beyond the boundaries of an isolated event. Yet what's equally interesting is how, as much as any recent California tremor, Loma Prieta occupies a territory of signs and symbols, an elusive middle distance in which interaction starts to merge with myth. What does it mean, after all, that both Allan Lindh and Jim Berkland called this temblor, that through the intercession of either precursory activity or earth tides and lunar cycles, some kind of progression was observed? I may have come away from my visit to Berkland extremely skeptical about his methods, but I have to give him credit where it's due. And Loma Prieta is one of the few earthquakes about which the geologists and the predictors agree on anything, even if they can't see eye to eye on how or why.

This nebulous interplay between science and intuition may be most visible in the Napa Valley town of Calistoga, about seventy-five miles north of San Francisco, and a twenty-minute drive from Berkland's house. There, surrounded by, as Thurston Clarke describes it, "fainting goats, wishing well, Ruritanian tollbooth, Vietnamese potbellied pigs, and mineral spring used for baptisms and hard-boiling eggs," sits a geyser called (what else?) Old Faithful, which, for the most part, erupts at intervals of forty minutes, shooting boiling water sixty feet into the air. In 1975, however, the geyser's owner, Olga Kolbek, noticed a lengthening in the lag time between eruptions prior to certain earthquakes, most notably a 6.1 that shook Oroville, in the Sierra Nevada foothills, on August 1. For nearly fifteen years, Kolbek kept detailed records of these variations, but it wasn't until Paul Silver and Nathalie J. Valette-Silver, then-married scientists from Washington, D.C.'s Carnegie Institute, visited Calistoga two months after Loma Prieta that anyone in the seismological community began to think seriously about what was going on. For the Silvers, the geyser appeared to have exhibited a "co-seismic response" with at least three substantial Northern California temblors—Oroville; a 6.2 that struck Morgan Hill on April 24, 1984; and Loma Prieta—each prefigured by changes in the eruption intervals, which returned to normal once the quakes had passed. "During the period 1973–91," the couple wrote in a 1992 paper published in the journal *Science*, "the interval between disruptions from a periodic geyser in northern California exhibited precursory variations 1 to 3 days before the three largest earthquakes within a 250-km radius of the geyser, including the Loma Prieta earthquake. Such precursive signals are one of the prerequisites for successful earthquake prediction."

The catch, of course, is that phrase "earthquake prediction," which remains as loaded a bit of language as exists in the seismic realm. Of all the geologists I've encountered, only a handful—Christopher Scholz (who as recently as March 1997 published an essay called "Whatever Happened to Earthquake Prediction?") and Allan Lindh among them—still use it with any regularity, and Lindh is just as likely to invoke the softer descriptive "forecast," which comes with less baggage, less association, less of a sense that we've crossed into a domain unknown. Because of this, the Silvers' work was greeted with derision by many seismologists, written off as a classic case of overreaching, of trying to forge connections out of coincidence. It didn't help that for more than a decade before their arrival in Calistoga, the most vocal proponent of the geyser-as-precursor theory had been Jim Berkland, who first met Kolbek in 1978 when they were introduced by a reporter from the *San Jose Mercury News*. ("Johnny-come-latelies," Berkland calls the Silvers, in his familiar scoffing tone.) Even the couple's explanation—that eruption intervals changed as a result of groundwater variations, which were themselves the product of preseismic activity leading up to the quakes—was dismissed as "fairly incredible" by the Bay Area Earthquake Task Force, because the tremors had happened so far from the Old Faithful site. "Since Loma Prieta's epicenter was 120 miles from the geyser," Thurston Clarke writes, "and since instruments closer to it had not recorded changes in the earth's crust, it seemed unlikely that this distant, hokey tourist attraction could be sounding an alarm."

It is here, however, that the fluid nature of geology once again asserts itself, with the most basic pieces of the seismological puzzle remaining adrift and floating, like tectonic plates of the mind. Shortly after the Silvers' research appeared in *Science*, the Landers earthquake established interaction as a seismic paradigm, and suddenly, the geographical distances separating Olga Kolbek's geyser from the epicenters of all those earthquakes didn't seem so inexplicable anymore. If earthquakes were able to trigger other earthquakes across hundreds of miles, why couldn't the same be true of precursors, as well? On the most basic level, the geyser's newfound status represented a vindication for someone like Berkland, evidence that, when it comes to earthquakes, truth occasionally emerges from the least expected corners. At the same time, geologists warned, Calistoga remained one small source of information, from which no real conclusions could be drawn. Just because the geyser was a pretty good seismic barometer did not prove the viability of prediction, or even precursory phenomena, *in and of itself*. As Paul Silver once told me, "It's really important to distinguish between quackery and science. These are observations only, and they don't say anything about the role precursors play as an earthquake develops, or correlate them with other precursors to get a sense of seismic strain."

In a certain sense, Silver's statement is the hallmark of a typically cautious scientist, one who doesn't want to say anything the data won't support. If we peel his phrasing back a little, however, we find ourselves returned to a landscape of complexity, of influence, in which Kolbek's geyser is just one of a wide array of fundamental processes at work within the earth. For all his use of the word *prediction,* what Silver's really after is something more like interaction, albeit a different level of interaction from the one Ross Stein describes. He is less concerned with connections between a set of earthquakes than with events leading up to a *particular* earthquake, the little burps and bubbles that indicate seismicity is on the rise. This may seem paradoxical, but like so much else in this ever-shifting universe, it resolves to an unanticipated unity once we open up our point of view. For Silver, it all goes back to Haicheng, with its foreshock sequence and well-water anomalies, its animal behavior and earthquake fog. None of these elements, he points out, mean anything on their own terms, yet taken together, they suggest a framework for approaching the Calistoga geyser—as a component in a sequence of preseismic interactions that mirror the larger interaction of which each earthquake is a part. The idea makes me visualize a three-dimensional cluster of fractals, as if I'm looking at a geologic Mandelbrot set. This is the name of the geometric phenomenon that has become, writes James Gleick, "a kind of public emblem for chaos," defined by endlessly repeating swirls and spirals, paisley patterns and seahorse tails. Among the Mandelbrot set's many peculiarities is that, the more you magnify it, the more it duplicates itself in increasingly miniaturized arrangements, not exactly (it *is* a chaotic system, after all), but within a remarkably consistent range. "The mathematicians," Gleick notes, "proved that any segment—no matter where, and no matter how small—would, when blown up by the computer microscope, reveal new molecules, each resembling the main set and yet not quite the same." In the face of that, I wonder whether it could be true of earthquakes also, whether the mechanics of interaction might exist equally on a macrocosmic and a microcosmic scale.

If there's any way to make sense of such a question, it is to look for other links, relationships—additional precursors that, when juxtaposed with Kolbek's geyser, lead us deeper into the overlapping layers of seismology and myth. Here again, Loma Prieta is full of surprises, especially in the area of water, and the extent to which it indicates activity underground. As things turn out, the Silvers were hardly the only geophysicists to uncover so-called hydrologic anomalies in conjunction with the earthquake; even before they arrived in California, the USGS's Evelyn Roeloffs was already looking into a report by Dan Friend, a one-time ranger at Big Basin Redwoods State Park near Santa Cruz, who had been hiking in the park just prior to the temblor when he noticed a significant increase in the flow of water over Berry Creek Falls. Roeloffs had long been interested in possible connections between seismicity and groundwater; among the patterns she'd investigated were potentially precursory variations in two of four monitored wells near Parkfield, beginning three days before the 6.1 Kettleman Hills earthquake of August 4, 1985. Friend's account, however, was somewhat different, if only because of its immediacy. In a 1993 USGS overview of preseismic observations, Roeloffs describes the situation:

> About 3/4 to 1 hour before the earthquake, Mr. Friend was at the observation deck below Berry Creek Falls. An abrupt increase in the sound of the falls attracted his attention. Over the next 4 to 5 minutes, he saw the flow over the falls increase to a final level that he estimated to be 3 to 5 times the original rate. Mr. Friend continued hiking above the falls, and he was setting up camp along Berry Creek just above its confluence with West Berry Creek when the earthquake occurred. Several large boulders came rolling downhill toward the campsite, and Berry Creek became turbid. Mr. Friend decided to hike out of the area immediately after the earthquake. On his way out toward the west boundary of the park, he noticed that streams which had been nearly dry earlier in the day were now flowing and that several new springs had appeared, including one or more in the hiking trail about 100 m below the

observation deck. West Berry Creek and Waddell Creek were both turbid after the earthquake.

The trouble with a report like Friend's is that contrary to, say, a regularly repeating geyser, it cannot be independently validated. It's a one-shot deal—now you see it, now you don't—and, as such, impossible to interpret on any but the most speculative terms. "I was very skeptical initially," Roeloffs admits one afternoon by phone from the USGS field office in Vancouver, Washington, where she's been based since 1991. "It introduces a lot of frustration. The typical scientific method is to experiment and repeat, and that was never an option here." Still, if this would seem to cast the events at Berry Creek squarely in the realm of anecdote, a couple of factors suggest that something more definitive may be at work. One is Friend's status as a former park employee, a man familiar with the area's topography and how it behaves. As a result, Roeloffs argues, his comments carry more weight than those of a casual visitor, who couldn't be expected to know the landscape with the same degree of nuance, nor to recognize anomalies. On top of that are Roeloffs's observations of the Berry Creek Falls area, made during two visits in November 1989. Although by then, the majority of activity had subsided, a few residual traces lingered, including boulders that had rolled down hillsides, and a number of large toppled trees. Most significant is what Roeloffs discovered when she retraced Friend's path along the creek bed—turbid water in West Berry Creek, moving at an abnormally high rate of flow. On its own, this may not conclusively prove Friend's story, but it does lend credibility to what he says. "The phenomenon that he describes," Roeloffs concludes, "is well documented to have occurred at least as an aftereffect of the earthquake. . . . Thus, all aspects of Mr. Friend's report that could have been verified after the event have been checked and found to be accurate."

For all Roeloffs's efforts to authenticate Friend's story, the most powerful evidence in its favor is ultimately how much sense it makes. It's not only the resonance between his experience and the Silvers' research that compels me, although this does make for a lovely bit of synchronicity, an impression of the mechanics of interaction as they move up and down the state. No, more to the point is the pattern of history, the way these incidents reflect both one another and a more extensive research record going back over eighty years. For much of the twentieth century, scientists have recognized a phenomenon known as fluid-induced seismicity, which suggests that as liquids ebb and flow within a region, they produce increased deformation, which, in turn, affects levels of seismic strain. Often, such a process begins with the extraction of oil or other fluids from subterranean reservoirs, although it can occur when water is pumped *into* the earth, as well. During the 1920s, writes geologist Paul Segall, several tremors struck near the Goose Creek oil field in South Texas after production there "caused the field to subside by as much as 1 m between 1917 and 1925." Similar observations were recorded in Long Beach throughout the 1940s and 1950s, while in the early 1970s, Colorado researchers flooded a derelict oil field to see if the ensuing change in water pressure would influence seismicity. (It did.) More recently, Roeloffs has gathered reports of spring flow variations prior to both the Landers and Northridge earthquakes, as well as information from Kobe, Japan, where for a month or more before the 6.9 temblor that killed five thousand people on January 17, 1995, increased water flow was measured on a continuous basis inside a mountain tunnel northeast of the city, from a stream that had broken through a cracked connecting wall.

What distinguishes the news from Kobe is that it's been documented in the official records, which means we're no longer trafficking in the elusive substance of anecdote, but have moved on to the hard, sharp blade of fact. By saying this, I don't mean facts alone have anything to tell us, just that *these* facts may help to collapse further the distances dividing science from myth. Kobe, after all, was an unanticipated earthquake; although Japan does have an active forecast network, the focus in the years before the disaster was on the Tokai region, which lies between Tokyo and Osaka in the southern part of the country, near the triple junction of the Pacific, Eurasian, and Philippine plates. Together, these plates form a subduction zone, with the Pacific pushing under the Philippine, and the Philippine doing the same to the Eurasian. Kobe, on the other hand, lies somewhat farther from the triple junction and, except for a 6.1 in 1916, had not suffered a major temblor since 1596. With an average interval of 1,000 to 1,500 years between events on this segment of the Arima-Takatsuki Tectonic Line (the fault that caused the tremor), it's no wonder the 1995 earthquake took everyone by surprise. "Kobe is like Japan's Memphis," Lucy Jones once told me, and it's a valid point. Memphis, too, notes Kerry Sieh, is "a candidate for large-scale seismic disaster," sitting in the hazard zone that emanates from the Reelfoot Fault; yet as is true of Kobe, it's a city with no earthquake culture, one that does not see itself in such apocalyptic terms. Then, of course, there's the confluence of anecdote and inquiry, the way that, like its Japanese counterpart, Memphis—the entire territory of New Madrid—occupies a landscape where these elements intersect. What's the difference between Susan Hough using modern methods to verify two-centuries-old newspapers and letters, and a set of water readings that, if accurate, support millennia of word of mouth? Nothing, I would argue, except a matter of degrees. Both suggest a relationship in which science does not so much contradict myth as emerge out of it—a framework, in other words, by which these ostensibly opposing concepts might fit together as a larger whole.

Of course, the closer we come to tracing a continuity between anecdote and science, the more we slip into an uncharted area where the layers of myth and investigation start to fold over on themselves. In many ways, it's not dissimilar to the "yin and yang" Ross Stein invokes, the idea that what's required is a sorting out. How do we identify a throughline when the terms are constantly shifting? At what point does conjecture spiral into fantasy, creating false connections that crumble at the slightest touch? These are the questions skeptics ask about interaction, with its essentially retrospective patterns—patterns that, Susan Hough contends, may even emerge in conjunction with New Madrid, since, if we look closely at accounts of the sequence, they indicate "a triggered earthquake the night after the third mainshock." With Kobe, however, we're facing a different situation, because the data there is forward, not backward, looking, a set of measurements gathered before the fact. That suggests a way of pushing interaction to the next level, of breaking through the boundary of the present and framing seismicity through a filter of future events. If we extend the concept back to Berry Creek Falls, and then again to Calistoga, we begin to see the edges of a whole new kind of thinking, one that may or may not tell us when a temblor will be coming, but at the very least, implies that *something's* happening, that there's a process at work, a system of arrangement, a structure by which, as anecdotal observers and predictors from Pliny through Cloud Man have long argued, the earth may send out signals in preparation for a major event.

Nowhere does this notion of signals make as bizarre an appearance as in the work of Tony Fraser-Smith, the Stanford University professor who recorded anomalous ultra-low-frequency (ULF) electromagnetic waves in the ground near Corralitos, a small town three miles from the Loma Prieta epicenter, in the two weeks leading up to that earthquake. Fraser-Smith's findings are significant for a couple of reasons, not least because their seismological applications are entirely accidental, which means they have a kind of serendipitous purity, bearing neither an agenda nor preconceptions about how the lithic landscape works. For much of the 1980s, actually, Fraser-Smith was funded by NASA, the National Science Foundation, and the United States Navy to measure naturally occurring electromagnetic signals, as well as seasonal and solar variations, to help in the planning of long-range communication with submarines and spacecraft, which often suffer interference from such disturbances. The experiment was essentially a hands-off operation, involving an unmanned monitoring station—"a house in the woods, with sensors"—and a computer hookup, by which information was collected once a month. Fraser-Smith didn't know he had anything until more than a week after the temblor, when a student gathering data caught the irregularities and called him in. "Twelve days before the earthquake," he says, "the noise level went up by a factor of ten. Three hours before, it went up by another factor of ten." Even so, Fraser-Smith's first reaction was to test his equipment, as if only a technical issue might account for what he'd found.

The discovery of what Fraser-Smith carefully allows "may have been magnetic precursors" to Loma Prieta represents the most explicit conflation yet between myth and science, because it opens up a passage not just to anecdote, but to the murky territory of the X-Files, the vague and subterranean landscape of the lunatic fringe. Once we start to talk about electromagnetic waves and ultra-low frequencies, we're back in the realm of earthquake sensitives, not to mention all those oddball theorists from Harvey Rice to Jack Coles to John J. Joyce. Shortly after Loma Prieta, in fact, Fraser-Smith accompanied USGS seismologist Andy Michael to Coles's San Jose apartment, where they found the predictor monitoring a symphony of static coming from an elaborate array of radios tuned between stations at the low end of the dial. "Clearly," Fraser-Smith remembers, "he believed in what he was doing. I don't think he was a charlatan. But every time a radio popped, he'd claim it indicated something, which he'd then interpret according to criteria that he wouldn't tell us anything about." For this reason, perhaps, Fraser-Smith's own electromagnetic measurements met with significant resistance when he first presented them to the seismological community, particularly at a December 1989 meeting of the American Geophysical Union, where he faced what he calls "tremendous bias, aggression actually" in defending his results. As is so often the case, however, this is an instance in which mainstream science operates out of far too narrow a focus, since what's important is not whether, in tracking these electromagnetic impulses, Fraser-Smith has lent credence to some elusive form of left-field thinking, but that he's come up with verifiable evidence of preseismic activity, be it precursory, or anomalous, or both. This is the equivalent of Kobe, another indication that as the earth prepares to rupture, it undergoes sudden stress changes from which a range of signals may arise. As to what that means . . . well, Fraser-Smith suggests, it all comes back to water, a resolution so circular, and so appropriate, I want to clap my hands. "The fact is," he says, "that for some reason or other, when fluids—and when I say fluids, I mean groundwater—are driven through porous rock, it's absolute solid science that you'll get a voltage generated between where the fluids begin and where they end. That's called a streaming potential. Once you have a potential in the earth, electric currents can flow through water and rock, and electric currents always generate a magnetic field, which is where my measurements come in."

One issue all this raises is whether Loma Prieta could have been predicted, legitimately predicted, whether the nexus of forecasts, interaction, and precursors might have operated as a set of portents, had anyone put together what was going on. That doesn't seem like an outrageous question, even for those of us who read seismicity as a language of uncertainty, a narrative of chaos theory inscribed deep within the ground. As Lynn Sykes, Bruce Shaw, and Christopher Scholz argue in their 1999 paper "Rethinking Earthquake Prediction": "Earthquakes, and the faults on which they occur are thought to be an example of a complex physical system that exhibits chaotic behavior. Yet such a characterization does not preclude useful predictions. . . . Chaos does not mean complete unpredictability or randomness, the common nonscientific usage of the word." By now, in any event, it's a moot point, another case of looking backwards, of piecing things together in the amber light of retrospect. Even if the data—anecdotal or otherwise—had been correlated, there's no way to know what we might have made of it, whether it would have yielded some kind of forecast, or simply added to the surface noise. Nor, unfortunately, can we say what this means for the future, because we're dealing with processes that, despite our best attempts to understand them, are sporadic, inconsistent, and keep analysis at bay. This is always the problem with prediction, the way it teases, tantalizes, offering something and then withdrawing it again. Tony Fraser-Smith's electromagnetic readings have not been replicated since Loma Prieta, just as the precursors that led up to Haicheng did not recur before the devastation of Tangshan. That leaves us in a murky middle ground of supposition, where the most documented evidence remains open to interpretation, and every query can be taken in at least two ways. "If you're measuring stream flow," asks Evelyn Roeloffs, "and you see something change, what do you do?" The answer is contingent on what happens, which is another way of saying that we may never really know just where we stand.

For the geologists, the solution is to measure strain directly, by going to the source. As much as water and electromagnetic variations compel us, they are, at best, what Paul Silver calls "indirect indicators," secondary signs and symbols, reflections of the slow and inexorable movements of the earth. Seismicity does not begin with them, but rather has its roots in the microscopic balancing and unbalancing of pressures that occur ten, fifteen, twenty kilometers underground. Because of that, Silver comments, "what we have to do now is enter the phase of really measuring deformation," a process made much more viable in recent years by the development of new technologies, including highly sensitive seismometers that can record small, previously untracable earthquakes, and global positioning satellites, which literally allow geologists to witness deformation across a broad geographic area by recording subtle shifts in the distances between sensors on the ground. Silver's sentiments are shared by researchers from Allan Lindh and Evelyn Roeloffs to Lucy Jones, for whom the value of such measurements has less to do with prediction than the more practical goal of hazard management, a standard by which, rather than concentrate so much on *when* an earthquake might be coming, we devote ourselves to identifying *where* it's likely to be. This is one of the possible applications of interaction, to use seismic patterns within a region as a way of providing better information about potential earthquake risk. "After Loma Prieta," Ross Stein says, "the Bay Bridge cracked and was Band-aided together, and we all know it's going to fall down again the next time an earthquake of that size occurs. Yet here we are, more than a decade later, and it hasn't been fixed in any lasting way. Why not? One reason is that we don't have data good enough to say the hazard is urgent, to say that we can't wait. If we could do a better job—coming up with more reliable numbers—the people responsible for public safety could make better decisions. For example, they might say, 'Give me a list of priorities. Which are the most hazardous faults from top to bottom? Tell us the likelihood of strong shaking on any one of these things.' Then, they'd know to pay attention to the structures built nearby."

It's hard to argue with the idea of making earthquakes safer, as anyone who's been through one will attest. Still, at the deepest level—and despite my own fears, my own worries—I can't help feeling that none of this is really to the point. For me, what interaction offers is less a way of living in the present than of getting inside geologic time. That's how I felt watching Stein's PowerPoint progressions, and it's how I feel also in the presence of these precursors, as if I've been enlarged beyond the limits of humanity, made part of a pattern that loops back endlessly upon itself. In such a landscape, there is no past, there is no future; there is just one continuous journey through eternity. And, like all equivalent journeys, this one has its own requirements, beginning with a leap of faith.

EAST OF EDEN

James Dean died in Parkfield. Or not in Parkfield, exactly, but fifteen miles south, at the intersection of California Highways 46 and 41 in Cholame, about twenty-five yards east of the Parkfield turnoff, which is Cholame Valley Road. In this spot, just before six o'clock on the evening of September 30, 1955, Dean slammed his Porsche Spyder into a Ford sedan driven by college student Donald Turnupseed, who was making a left from the eastbound lane of 46 onto 41. As the last minute of his life unfolded, writes David Dalton in *James Dean: The Mutant King*, Dean watched Turnupseed start to slip across the center line, then turned to his traveling companion, a mechanic named Rolf Wutherich, and said, "That guy up there's gotta stop; he'll see us." In the gloaming, however, Dean's low-slung silver sports car blended with the horizon, leaving Turnupseed oblivious to the forces propagating towards him until it was too late. "I didn't see him," the twenty-three-year-old cried at the scene of the accident. "I swear I didn't see him." Ultimately, then—as, in some odd way, seems only appropriate—the story of Dean's death begins and ends with a trick of the light.

Here is something else, though: at that same intersection—or, more accurately, ten to fifteen kilometers beneath it—seismologists now believe the Fort Tejon earthquake began its rupture, nearly a hundred years before Dean embarked upon his final ride. This is my absolute favorite earthquake fact ever, that James Dean died at the Fort Tejon epicenter, as if the location itself, the very crossroads, might be a kind of vortex, a magnet for our most enduring myths. I don't want to suggest that Dean and Fort Tejon have anything to do with each other, although were I to look hard enough, I'm sure, I could discover someone who would tell me that they do. But I'd be lying if I said I didn't find the coincidence compelling, if only because it means that out here, among the crumpled hills and scrub flats of Central California, all the state's strange symbology—speed, youth, celebrity, and earthquakes—comes together, making this a place that speaks to us on any number of levels all at once.

Highway 46 is a ninety-mile ribbon of two-lane blacktop that unrolls out of the San Joaquin Valley, from Wasco through Cholame to Paso Robles, along the northern edge of the Carrizo Plain. Here, the towns have names like Lost Hills and Devils Den, and even early on a Tuesday morning in November, the sun is hot, a floodlight in the middle of a sky as white as summer; there are no clouds. As I push west, I watch the topography change in increments, from dust-strewn desert to citrus orchards to the dull brown sheen of the Cholame Hills. Every now and then, I see cows grazing in clusters behind fences, while to the south, the slate gray moonscape of the Temblor Range rises like a set of incisors, casting jagged silhouettes against the morning light. It's a forbidding landscape, barren, elemental, and, in truth, the isolation makes me feel a little nervous, like I'm on the verge of something that I can't control. At the same time, my anxiety keeps giving way to something else—an edge of anticipation, of exhilaration—as if, in coming to Cholame, I have driven into the middle of a vast blank canvas, a leaping-off point, a template for our fantasies and our dreams.

The irony is that I'm in Cholame today more or less by accident, that I am only passing through. My destination, after all, is Parkfield, but James Dean's ghost keeps getting in the way. Last week, when I called Andy Michael for directions, he mentioned 46 and Cholame Valley Road, then added that if I found myself at the James Dean Memorial, I'd have gone half a mile too far. On the phone, it sounded simple enough, yet the road to Parkfield is poorly marked, and I end up driving past it, continuing west until I reach the Jack Ranch Café. This one-story log cabin is the site of the Dean Memorial, an angular chrome sculpture erected in the middle of a gravel parking lot around the trunk of a large tree. Inside the café, a few silent diners eat their breakfasts beneath a series of roughly drawn Dean portraits, which hang from every wall like icons in a church. When I ask the cashier about Parkfield, she repeats Michael's directions, assuring me that, from this side anyway, the route is pretty clear. On my way to the car, I wander over to the sculpture, which sits set off by a low divider built of four-by-fours, with two metal plaques, each bearing a date of dedication and a few phrases of remembrance, worn to near-indeterminacy in places by twenty years of wind and sun. Were I asked, I'd probably call myself a Dean fan; I've seen his movies, and in high school I kept his picture on my wall. I've even been *here* before, detouring once on a drive from Monterey just to see it, to stand at this spot and take it in. That was in 1994, a few months after Northridge, and as I recall, there was another plaque, featuring an Antoine de Saint-Exupéry quote that Dean is said to have favored as an epitaph: "What is essential is invisible to the eye." This morning, though, I can't find Saint-Exupéry's phrase anywhere, which makes me wonder about the line between memory and imagination, the difference between what we see and what we believe. The feeling is heightened by the scattered offerings laid out at the base of the memorial—a spray of roses still wrapped in flower-shop plastic, and a dusty pack of Camels rubber-banded to a longneck bottle of Bud. On the sculpture is the simple designation, "James Dean 1931 Feb 8–1955 Sep 30 ∞." The "s" in "James" and the "8" in "Feb 8" have been pried free, leaving little more than spectral traces of themselves.

As I stand there, the desolation of the area creeps back over me, and I feel an involuntary chill. A slight breeze has begun to blow north off the Carrizo, but it's no comfort, just a whistling emptiness across the static surface of the road. All around, I have the sense that things are in suspension, that time has slowed to liquid honey, that here is a place where I might reach out and touch both past and future, where the boundaries have gone thin. This, of course, is another intimation of eternity, not unlike what I've experienced with Ross Stein and Sally McGill. Yet today, eternity feels different—less theoretical, not so much a matter of fluidity as a state in which everything that has ever happened and everything that will ever happen coexist in a never-ending present tense. For a moment, I can almost make myself believe it, almost feel the lines of history slip, taking me backwards and forwards all at once. What, I wonder, will I see next on this highway? Will it be a silver Porsche Spyder heading for a Ford sedan? Will I hear the screech of tires, the metallic roar of impact? And how about the earthquake? Could that be coming also? A rumbling underneath me like the 1857 tremor come to life? I squint my eyes and peer into the distance, as if anything could happen, as if all this, too, might emerge out of the light.

Then an eighteen-wheeler passes, its trailer bed piled high with fresh-cut hay bales, and the concussion of air it causes knocks the moment loose. As a scattered string of cars and ranch trucks follow, the silence yields piece by piece to road noise, the hiss of tires, the thrum of motors, the Dopplered rhythms of radio bass and drums. I stay at the monument for another minute, and reach out my hand to trace the missing letter in James Dean's name. Finally, I get in the car and drive the thousand or so yards east to Cholame Valley Road, where I wait for a break in the oncoming traffic, before I turn left and inch slowly across the center line of Highway 46 like some latter-day Donald Turnupseed.

Parkfield is a town caught out of time. Or perhaps it would be more honest to say that it exists within its own time, a time both geologic and human, in which the minutes are marked primarily in terms of waiting, waiting for the earthquake that never comes. The whole place has been transformed by such a process, yet in another way, barely transformed at all. If this seems like one more contradiction, it might be the contradiction most fundamental to the way we live in California, which is why I've traveled here this morning, to find out if I can trace a pattern of reconciliation in a landscape suspended between myth and science, along the hardscrabble incision of the San Andreas Fault.

The tricky thing about the San Andreas is that, after a century of notoriety, it exists in equal measure as topography and symbol, by turns a physical and psychological dividing line. For all that it may cut (to return to Kerry Sieh and Simon LeVay's description) a "dramatic gash," an "infamous fracture," I can count on one hand the people I know who have actually seen it—except, that is, for those in the seismological community. Instead, despite its looming omnipresence, the fault continues to occupy for most of us the territory of imagination, like a giant geological bogeyman. Above my desk, beside a shake map of the Northridge earthquake, hangs an enhanced photograph, taken from the space shuttle *Endeavour*, of the San Andreas as it runs along the San Gabriel Mountains near Palmdale. In the center of the shot, you can see the fault, see the scarps and ridges as they rise on either side of it, see the terrain spell itself out like some strange runic language, a hieroglyph of indeterminacy. Still, although I study that picture often, stare at it, touch it, even run my finger down the image of the fault trace, there's a way in which it blurs into abstraction, revealing less than meets the eye. I love earthquake photos, love to ponder them, to dream my way beneath their two-dimensional surfaces. But ultimately, if my investigations have taught me anything, it's that a photo can tell us only so much, after all.

In Parkfield, though, I don't need to think in terms of photographs, because the San Andreas is everywhere. It's the first thing I see upon arrival and the last I leave behind me when I go. From Cholame Valley Road, I turn right and literally drive straight into it, a winding serpentine that unfurls beneath a one-lane bridge reinforced with concrete pylons, where a battered metal sign announces: "San Andreas Fault. Now Entering North American Plate." Here, the San Andreas resembles nothing so much as a dry riverbed, about a hundred feet wide, covered in gravel, lichen, mud, and bits of brush, with the merest trickle of water cutting a dirty chasm through the rocks and loose debris. This is utterly unlike the way it looks in Redlands—wider, flatter, less . . . well, *nuanced*—but as I leave the car, I recognize the same pervasive air of silent awe. Immediately, I want to sink into the fault, to immerse in it; although an old fence marks the edge of the rift (wooden posts wrapped in rusty barbed wire, "No Hunting or Trespassing" signs hung like empty warnings), it's so busted up that I have no trouble slipping through a gap and climbing down a couple of feet. There, something stops me from going further, a nearly physical edge of apprehension, the sense that I'm about to cross a line. I take a deep breath and feel the landscape swell around me, feel the sheer weight of its possibility, its uncertainty, of the idea that I am at the seismic precipice, where everything begins and ends. In the distance, beyond the sun-seared hills, I hear the low drone of a propeller plane, while somewhere in the sky's vast whiteness, birds chirp like tiny reassuring ghosts. Yet for all that this helps keep me rooted in the present (*three hours, no earthquake for at least three hours*), I can't help noticing how the present fades into an afterthought before the eternal stillness of the San Andreas, a stillness that exists beyond either myth or science, as absolute and ineluctable as all those forces sliding in and out of balance deep within the earth.

The notion that I am hovering at the brink of infinity lingers even after I return to my car and cross the bridge into Parkfield proper, which is a town, it turns out, only in the most vestigial sense. As I drive in, I see another sign—

PARKFIELD
POP. 37 ELEV. 1530

—a population so ridiculously tiny, I have to stop and read it twice. This is the smallest community I've ever been in: one narrow road, no more than four or five blocks long, fronted by an abbreviated cluster of single-story houses and a gift shop called the Parkfield Log Company, which is closed. Also closed are the

Parkfield Inn and the Parkfield Café, a matched set of rough wood structures, each dressed up with rusty planters made of old train parts, on which have been painted, in fake old-timey letters, variations on the slogan, "Earthquake Capital of the World. Parkfield, California. Be Here When It Happens." Past the café, there is a row of mailboxes, and beyond that, a gazebo; at the far end of town, just before the road arches back across the San Andreas, a derelict railroad siding holds a vintage Santa Fe freight car and caboose. In the other direction, a trailer doubles as a library, while a low, flat building serves as a one-room schoolhouse, no more than five hundred feet from the fault. A handful of kids run around the dusty playground, screaming and chasing one another, but amid the immensity, their voices sound as thin as whispers, and the whole scene only becomes stranger when I realize no adults appear to be around. Even the USGS monitoring station, a new-looking trailer set back from the road, looks empty and shuttered, as silent as the San Andreas itself. Although many of these buildings have clearly ridden out their share of earthquakes, Parkfield feels as ephemeral and two-dimensional as a stage set, as if the only thing that isn't temporary is the fault.

If Parkfield has any particular message to offer, it's that the earth outlasts us, that the issues of influence and strain and interaction will continue to play out upon this landscape long after we, with all our petty vanities, have disappeared. In that regard, Parkfield may be most important as a microcosm, a metaphor, not for earthquakes—not exactly—but for how we live in their proximity, the double vision they require, the delicate pas de deux. I'm not saying that we should read Parkfield through the filter of illusion; there is nothing illusory about the San Andreas, which holds the town in its embrace like some ancient ancestor, a spirit in the ground. No, what makes Parkfield compelling is precisely the opposite, that it occupies a territory outside illusion, that even to visit for one hour is to experience a peculiar clarity, a visceral appreciation of *the way things are*. Such clarity can elude you in Los Angeles or San Francisco, where it's surprisingly easy to keep seismicity at a distance—as my friends' unfamiliarity with the fault goes to show. When the San Andreas runs through your backyard, however, you have no choice but to stare down the precariousness of existence on a daily basis, "to accept, consciously or unconsciously," in the words of Joan Didion, "a deeply mechanistic view of human nature." Didion was referring to the Santa Ana winds when she wrote that, but she might as well have been discussing Parkfield, for what she's saying (and what Parkfield means to tell us, also) is that in California, we dwell less in harmony with nature than at its discretion. To ignore this is to ignore not just the risk of living in a fault zone, but the very nature of our place here, the precariousness of existence in an ever-changing world.

Of course, it's one thing to accept "a deeply mechanistic view of human nature," and another altogether to say what this might mean. We may be at nature's mercy, but especially when it comes to earthquakes, the mechanisms are elliptical, elusive, revealing themselves through hints and suppositions as much as any process we can see. This is the essential paradox, the one to which I keep returning—that, even in the shadow of the San Andreas, seismicity confounds us, operating at a level forever slightly out of reach. Whatever else it may represent, after all, Parkfield is best known as the town where the USGS took a gamble on prediction and lost. That's the public perception, and to move through this deserted landscape is to feel the weight of such a sentiment, a weight that is, in many ways, as heavy as the fault itself. In the late 1970s and early 1980s, when Allan Lindh and others first set up shop, they must have brought a real air of discovery, of history in the making, the idea that Parkfield's time had come. Two decades later, what remains are slogans: "Be Here When It Happens," "Earthquake Capital of the World." There's something oddly optimistic about these dated come-ons, with their assurance not just that anything could happen, but that it *would*. The one result nobody counted on, though, was silence, which tells us how naive everyone was. Here, we find the second lesson of Parkfield, the human lesson, the one that says things are more complex than they appear. This, in turn, brings us back to the narrowness of the seismic model, the way six temblors in 150 years look increasingly inconclusive the more you open up your sense of time. How can we say what these quakes mean if we don't know the history that precedes them, let alone the record of what's to come? It's even possible that the entire sequence could be anomalous—or, as Andy Michael puts it, "Maybe these six earthquakes aren't a periodic set. People interested in chaos theory point out that even chaotic systems will occasionally do things in what looks like a regular pattern, until it breaks down and they do something else."

Michael means his comment as an exercise in conjecture; a self-described prediction "moderate" who was, for many years, acting chief scientist at Parkfield, he likes to play with possibilities, admitting that he doesn't always know just what he thinks. All the same, he reminds me of the intricate dance of time that we must do in Parkfield, where human chronology constantly telescopes into geologic and back again. Such an idea is only complicated by an additional overlay—the absolute, inevitable certainty that, for all the vagaries of frequency and magnitude, the San Andreas Fault will one day slip again. This is the ultimate nuance, the wild card, the key piece of (unknown) information, and it means that Parkfield remains very much a work in progress, on the experimental as well as the experiential plain.

Public opinion notwithstanding, the USGS has never given up on wanting to instrument a Parkfield tremor, and research efforts are still ongoing, from strain meter experiments that, Michael explains, "measure deformation to tell if the ground is being compressed or stretched or sheared" to the more observational studies of Evelyn Roeloffs and Tony Fraser-Smith, both of whom have had equipment running at Parkfield for years. If you talk to Lindh or Michael, they'll point to the variety of investigations—"the density of the array," Michael calls it, in a particularly lovely turn of phrase—as the experiment's most significant achievement, because it has allowed the USGS to build a seismic network, which, when the next earthquake rolls through, should provide the kind of firsthand data that, until now, has been all but impossible to record. You can't visit any of these experiments; much of the machinery has been installed on private property, or lies buried underneath the ground. But this, I think, is exactly as it should be, for Parkfield is a town where the most important story has always taken place below the surface, where, indeed, "what is essential is invisible to the eye."

What we make of this depends upon our outlook, the ideas we hope to reconcile. At the most pragmatic level, it means that Parkfield is as close as we've come to the direct study of deformation Paul Silver talks about, a broadly monitored landscape where different strands of research may be woven into the tapestry of a larger point of view. I'm especially compelled by the fact that the work here combines instrumented and anecdotal science, that there is room for everything, from seismographs to water wells and electromagnetic fields. This, I'd argue, is the only path to understanding, to circle the fault and see what happens, as if in the subtle interplay of measurements and precursory activity, we might find a passage back to geopoetry. Thinking about it, I remember that, as Allan Lindh has pointed out, short-term prediction was never intended to be more than a single piece of the experimental puzzle; equally important was the concept of woolgathering, of collecting data, of waiting for the earth to assert itself, and then identifying whatever details might emerge. This is the final thing Parkfield has to teach us, the notion that, in a landscape bounded on one side by eternity and on the other by complexity, time itself is always in suspension, stretched simultaneously between the present, the future, and the past. It's been twenty-three years, after all, since the USGS arrived in Parkfield, which is a span more extensive than my adult life, longer than my marriage or my oldest friendship, going back further than the ages of my kids. On the fault line, however, twenty-three years is less than nothing, an instant so infinitesimal that we cannot say anything definitive about it—even whether or not the prediction here has really failed. "The interesting thing," Andy Michael notes, "is that depending on what statistical distribution is used, the Parkfield earthquakes still look sort of periodic, periodic with error, until about 2003. Somewhere between 2001 and 2003, depending on the test you use."

Consider Michael's analysis one way, and it sounds like nothing so much as a hedge, a manipulation of the data, the kind of statement someone like Jim Berkland might make. How could a forecast originally meant for 1988, give or take five years, still apply to a temblor ten or even fifteen years later, statistical distributions aside? Look a little deeper, however, and his numbers start to get at how wide open and undefined the territory of prediction is, with its influences and interactions that double back and forth across one another in an endless game of seismic telephone. After all, whatever we think about the so-called Parkfield earthquake sequence—whether we accept it as indicative of something, or see it as far too limited a sample—we can't avoid framing it in terms of patterns, and patterns, as Ross Stein and Kerry Sieh have demonstrated, are tricky things. Even if we restrict ourselves to the six recorded Parkfield temblors, we come up against discrepancies: although the average interval between events is twenty-two years, not one gap in the sequence has ever hit that total, and the two most recent tremors, those of 1934 and 1966, are separated by an entire decade more. When we add thirty-two years to 1966, we get 1998, which, with a plus-or-minus-five-year margin of error, brings us up to 2003. Then, of course, there's the issue of chaos, the possibility that the whole progression may be random, or, at least, subject to the effects of other activity in the region, like the 1983 Coalinga quake. Whether or not this means the next Parkfield earthquake is imminent, I really couldn't say. But it does suggest that, once again, we are adrift in a floating universe, where neither science nor myth may be sufficient to account for everything we do not know.

All these thoughts tangle up inside me as I make my way out of Parkfield, taking the slow sweeping curve of road back to the San Andreas bridge. On this side, the sign that greets me is the opposite of its partner: "San Andreas Fault. Now Entering Pacific Plate." Briefly, I smile at the uninflected matter-of-factness, the way it asks not where nor how, but simply lets us know what is. Then, I inch my car onto the bridge until I am directly in the center, where I stop and turn off the engine, suspended now not just in time but also in geology, halfway across a fault that rests beneath me like a silent river, locked and waiting to resume its flow.

I don't stay on the bridge for very long, maybe a minute, maybe two, and I don't get out of the car either, in case another vehicle comes along. Nevertheless, there's something about being here that gets inside me, that makes me feel like I am sitting at the center of the world. Through my window, I watch the San

Andreas wind through the fields and hillocks; among the brush that grows out of its stirred-up soil are occasional clusters of small trees, and I can see their stunted crowns peek above the edges of the chasm in a thin green line, marking a passage south, towards Cholame. It is in this direction that the 1857 Parkfield earthquakes propagated—speeding, like James Dean's silver Spyder, towards that final intersection, where they helped trigger Fort Tejon—and as I look along the fault trace, I find myself staring into eternity again. What if the next Parkfield temblor were to rupture right now, I think; what if Andy Michael is right, and I am sitting on the cusp of a seismic window, at the exact instant geologic and human time collide? The idea gives me a thrill, a jolt of excitement: not that I *want* it to happen, but it would make for the earthquake story to end all earthquake stories if I were here when the tremor hit. Quickly, I try to calculate the odds, try to imagine what it would be like, the bridge rocking, the guardrails bending, the very earth rearranging itself in a flowing symmetry of curves. It's a captivating image, and in its sway, I am momentarily transported—until, that is, I recall the birds, and the sound of their singing, which tells me that whatever earthquake may or may not be coming is at least a couple of hours away.

On the drive back to Los Angeles from Parkfield, I avoid the freeway, following a more elliptical path (is there any other kind?) along the San Andreas, across the starkly desolate expanse of the Carrizo Plain. From Cholame, I take 46 east to Blackwell's Corners, where James Dean made his final stop less than an hour before he died. There, I turn south on Highway 33, another two-lane strip of road that runs parallel to the California Aqueduct, just east of the Temblor Range. To my left, farm fields lie fallow, brown and dusty, while on the right, oil wells glint in the sunlight, pumping up and down like prehistoric birds. The farther south I go, the more these wells multiply—until, near the town of McKittrick (population 190, with a school, a market, and an old redbrick hotel), hundreds, thousands, tens of thousands of them stretch back from both sides of the highway, a metropolis of production, with massive pipelines that extend along the roadbed, and signs identifying this as a joint venture of Chevron and Texaco. Although the San Andreas runs twenty miles to the east, there are other, more immediate seismic issues; according to a field guide published in 1994 by the Department of the Interior's Bureau of Land Management, this is a "region of moderate to intense folding and faulting . . . [in which the] major subsurface structures are the Cymric and Westrich anticlines, which are separated by the Bacon Hills Fault. The Cymric anticline is a doubly plunging, northwest-trending subsurface structure cut along its apex by a network of steeply dipping faults." That faulting and folding, of course, is responsible for the oil deposits, but I wonder to what extent the drilling may have pushed this area closer to failure, amping up stress on all these intersecting fault lines, a delicate cat's cradle of shear and influence that could, at any moment, slip out of control. In 1857, the last time a great earthquake shuddered through here, the earliest oil production was thirty years in the future, and yet, the aftereffects of that tremor remain visible across the landscape, in the form of countless mud-slid canyons and shattered trees. Now, I stare out across the roaring tumult of machinery, and consider what a good-sized quake might do.

Once I pass through McKittrick, I turn west on Highway 58 and move into the Temblors. Very quickly, I am alone in the universe; there are no cars, not a house nor a barn nor a pasture, not even the hypnotic davening of an oil well. As I drive, the road rises in a series of terraced switchbacks etched into the sides of crumbly mountains; everywhere are traces of alluvial fall. The terrain is scarred with great gaping fissures and crevasses, all looking like their own fault segments, as if what I'm traversing is less a piece of geography than some enormous jigsaw puzzle, strewn haphazardly in pieces on the ground. Finally, after fifteen miles, the highway descends along a deep ravine and out onto the broad, flat plain of the Carrizo, a territory as vast and empty as eternity itself. In the distance, haze shrouds the Sierra Madre, while sunlight glitters off the receding crescent of Soda Lake. I slow to a near-crawl, and take a left on Seven Mile Road, where a weather-beaten structure stands shuttered in the November sun. This is the Carrizo Plain Store, and the San Andreas slices just in front of it, crossing beneath the road at a right angle, before continuing south to Wallace Creek. It was at Wallace Creek that, more than a quarter century ago, Kerry Sieh first began to investigate the San Andreas, and, in the process, invented paleoseismology. And it is at Wallace Creek that we can still find an especially vivid illustration of what geologists call an "offset channel"—for where the creek once cut a straight channel east to west out of the Temblor foothills, there now exists a 420-foot north-south dogleg, caused by thirty-eight hundred years of earthquakes producing slip upon the fault.

To get to Wallace Creek, I take Elkhorn Road from the Carrizo Plain Store and follow it four miles into the abyss. Four miles sounds like a quick jaunt, a negligible distance, but it might as well be four hundred or four million, because out here, I'm in as alien a world as I have ever seen. Elkhorn is an unpaved road, deeply rutted, covered with rocks and a fine white dust like ash. As I bounce along, I feel like I am moving counterclockwise, into a time before time, before chronology, as if I had traveled back to the Precambrian, to some uncharted period of prehistory where silence is the only logic that applies. Even the few human traces

I find appear to exist outside any order, speaking to me of untamed edges, of a psychic region I don't recognize. Half a mile before Wallace Creek, I pass a shattered camper shell rising derelict and discarded out of a clearing thirty feet from the road, while next to it, a red Volvo lies on its side, doors open and engine gutted, like a carcass picked apart by carrion. I get scared when I see that—truly frightened—as if, for the first time today, I am in the presence of something that could do me harm. This may sound strange, given what I've spent my morning doing, but the truth is that I'd rather take my chances with the San Andreas than with whoever abandoned these vehicles like so many forgotten playthings; I'm alone, after all, and no one knows exactly where I am. In *Magnitude 8*, Philip Fradkin describes a similarly disturbing discovery, along the bend of Wallace Creek itself. "There was more grass," he writes,

> albeit skimpy, at one spot; and sprinkled about were bits of charcoal. A man-made fire had fertilized the pebbly stream bed. . . . Among the small chunks of burnt wood, I found whitened bones that appeared to be from an avian creature, a rabbit, and a cow or lamb. Mixed in was the remnant of a red 1995 registration sticker from a California license plate. A flat piece of burnt material, perhaps plastic, was calcified and pierced by a rivet.

What it all adds up to is a forsaken landscape, not just barren but apocalyptic, a place you land when there is nowhere else to go. As I park and hike to the creek site, I have the overwhelming sense that I could get lost here, that I could take a wrong turn, or meet the wrong person, and literally never find my way home.

Wallace Creek sits about a hundred yards east of Elkhorn Road, tucked into a fold between two low rises, a long trench cut into the Carrizo floor. At first, I'm struck by how much it resembles the fault in Parkfield: the same gray mix of dirt and gravel, the same sporadic scrub brush, the same sensation of deep and lasting stillness, as if I had stepped inside a moment that will never end. It's not nearly as wide, however—maybe fifteen feet, maybe twenty—and although you'd never know it from the surface, the fault is visible only in the dogleg, emerging briefly before continuing, beneath the hills and pressure ridges, on its jagged, winding passage to the Salton Sea. Still, as much or even more than Parkfield, Wallace Creek is a place where seismicity is right in front of us; I can see it as clearly as an X-ray, a record of how the earth has changed. To the north, the original stream channel sketches a faint trace on the foothills, and as I move my gaze south, panning movie-camera slow, it's like watching four thousand years unfold. The shift is as implacable as geology, and in the face of it, I catch a glimpse of what Kerry Sieh must have felt on his first visit, that wide-open edge of possibility, of time expanding outward, the sense that *everything* is in constant metamorphosis, which renders our most carefully constructed definitions moot. It's this blurring of boundaries that earthquakes offer, if we can figure out how to read them, and never more than when we appear before the evidence of their power. Yes, that evidence tells us, you could disappear here, but you might also be enlarged beyond your wildest imaginings, joined not only to the world around you, but in some fundamental fashion, to yourself.

There's a way in which this all gets back to the idea of life as mechanistic, as subject to the intercession of great forces, forces beyond our control. How, after all, can I deny such an influence when I'm currently in a spot that shifted thirty-one feet during the Fort Tejon quake alone? Thirty-one feet is the width of my house; it's a bus length, a first down, the distance from that camper shell to Elkhorn Road. Thirty-one feet will kill you, if you're in the wrong place when it happens, just as Sally McGill suggested that it would. Yet thirty-one feet—or, by extension, 420—is also a doorway, one that opens up the universe, allowing me to feel these forces, observe them, get inside them, to know them on the most elemental terms. As at Parkfield, this is not about illusion: no, in Wallace Creek, it's impossible *not* to recognize a progression, not to see the way that earthquakes intersect. The sensation only deepens once I drop down into the creek and start to walk the offset, moving slowly, stopping every few seconds, trying to feel the energy that (for the moment, anyway) lies dormant in the ground. Although I look, I don't see Fradkin's bones, or any fire residue—in fact, submerged as I am behind the creek's embankment, I don't see much at all. Again, I have the sense that I'm the only living creature in the universe, that were I to climb out of this streambed, I would find nothing, no trace of human existence, not even my own car. It's a desolate sensation, but different, more associative, than my earlier isolation, and reassuring in an unexpected way. To be alone here, after all, means that, on some essential level, I am safe, immune to anyone else's intentions, that I have crossed over, in other words, into geologic time, immersed myself within the seismic pattern, found communion with the San Andreas Fault.

Of course, it's hard to talk about communion without getting into the idea of God, of spirit, which is not exactly what I have in mind. At the same time, I'd be lying if I said I didn't feel the presence of *something*—call it faith, or legend, or geology; call it what you will. "I'm living in the Bible's world right now," Denis Johnson once wrote of Somalia, "the world of cripples and monsters and desperate hope in a mad God, world of exile and impotence and the waiting, the waiting, the waiting. A world of miracles and deliverance, too." The same is true of the Carrizo, which is, in many ways, a territory east of Eden, in both human and geologic terms. There are monsters here, both real and metaphorical. There are those cars, and whoever destroyed them, not to mention bone-burning rituals in the desert, all of which reflect the wild, dark side of California, the California of Donald Dowdy—who I can easily imagine in this primal landscape, stomping his foot "on the Andreas rift" to invoke an earthquake—of Charles Manson and Bishop Pike. There is exile (what else, really, is this place about but exile?), and as always, there is the waiting, all the endless waiting, for the moment that the fault will finally go. Still, if the Carrizo has anything to tell us, it's that seismicity offers miracles also, starting with the dogleg in Wallace Creek. To the naked eye, it's the most prosaic sort of wonder: nothing like the dream of a collapsing building, nothing that makes us stop and catch our breath. It's a miracle we can understand, a miracle based on constant repetition, but if that brings it nearer to the realm of science, it also reconciles the vision of a larger order, which is the most quintessential kind of earthquake myth. I want to be careful about what I say here, because that offset channel is not a matter of superstition, but an observable phenomenon, and yet it goes beyond science in what it shows of how seismicity leaves its mark. It's in the sweep of time, of eternity, the way that every day, in this location, the history of the San Andreas comes to life. That may or may not be a mystical experience, but I'd call it a miracle all the same.

I walk out along the creek bed for a couple of minutes, then come back to the dogleg one more time. Again, I wonder what it would be like if there was an earthquake, how it would affect me, whether I'd survive. According to Kerry Sieh, the average interval between major events on this segment of the fault is 160 years, which means we're not due until 2017, but that's just a statistic, and says nothing about when a quake might really come.

To find out, you'd have to ask the San Andreas, and the San Andreas keeps its secrets close.

By now, it's midafternoon, and the sun has gone all thin and reedy, like a candle burning towards the nub. I feel the faint chill of encroaching evening, and start to think about heading back. I want to be off the Carrizo by nightfall; I have no desire to wind up lost or exiled, nor to fall prey, like James Dean, to a fleeting figment of the light. Still, I can't shake the idea that something is going on here, even (or especially) if the ground remains unmoved. To stand inside the San Andreas is like balancing on a line between control and chaos, the opposite of, say, the earthquake ride at Universal, an experience that isn't manufactured, that contains every possibility, even that of nothing taking place. What happens next is, in the most acute sense, unpredictable; I won't—*can't*—know it until it comes. This, too, represents its own small miracle, a miracle of a particularly open-ended kind. I think about an earthquake propagating down from Cholame. I think about the Parkfield foreshocks. In 1857, that is how it happened, and in this empty landscape, it feels as close to me as my own breathing: the collapse of time, the eclipse of history, the Fort Tejon temblor rumbling again. I take a seat on the creek bed, feel a slight buzz of anticipation. And then, before I know it, I am lying full out in the fault zone, my heart pounding a rapid drumbeat, as for this one quick sliver of eternity, I set aside my hopes, my fears, my very identity, and wait, wait, wait to be delivered by an earthquake that does not arrive.

THE UNIFIED FIELD THEORY OF EVERYTHING

The Hector Mine earthquake arrived while I was sleeping. It found me in bed, blissfully unfocused, deep in dreams. I can't recollect the dreams anymore, only that the quake insinuated itself as a kind of ululation, less a matter of feeling than of sound. When the noise grew loud enough, I awoke, sitting up in bed with a start. I looked at the clock on the nightstand: it was 2:46 a.m., Saturday, October 16, 1999, one day short of the ten-year anniversary of Loma Prieta.

I want to tell you that I responded in some way to this situation. I want to tell you that I leapt up, that I went to get my daughter and my son. I want to tell you that there's a story to this temblor, that there's an obstacle to be negotiated, a bit of detail I cannot forget. The truth, however, is that nothing so dramatic happened, that I just sat there, rocking in the tremor's lazy oscillations, which, even then, felt like they were coming from far away. The movement was so slight that I wrote it off as a minor earthquake, a 3.5 or a 4.1, the kind that disappeared almost as soon as it got started, more of a whisper than a scream. (A year or two later, when I mentioned this to Lucy Jones, she laughed and told me, "Oh no, it went on for much too long.") My wife, Rae, had gone to Philadelphia for a family wedding, so once the house stopped shaking, I slid out of bed and looked in on my children, neither one of whom had stirred. I turned on the hall light to check the electricity, then picked up the telephone and heard a dial tone. No big deal, I recall thinking, as I padded back to my bedroom. Just another night in California, just another night of living in the earthquake zone.

The first clue that I was wrong came at about six a.m., when the phone began to ring. I woke up, looked at the clock, thought about answering, rolled over instead. Two minutes later, it rang once more, and, when I let it go a second time, two minutes after that. I got up, moving, as if in a dream, through the gray half-light of dawn. Vaguely, I worried about an emergency. Why would anyone call this early on a Saturday, at an hour when the universe itself was sleeping, unless there'd been an accident, a death in the family, some disastrous circumstance beyond either expectation or control? It never occurred to me that the circumstance in question might be the earthquake—to tell the truth, I'm not even sure I *remembered* the earthquake, so small and insignificant did it seem. Then, I picked up the phone, and on the other end was Rae, a metallic presence in the handset, her voice run through with a fine, sharp edge of panic, as piercing as the point of a blade.

"Are you okay?" she asked, no preamble, no good morning, no indication whatsoever of why she'd called.

"What?" I said, my brain dense with confusion. "What are you doing? It's six in the morning."

"Are you okay?" she asked again.

"I'm fine." I said. And, when she didn't answer: "Why? What's the problem? Is something wrong?"

"There was an earthquake . . ." she said, each word a tiny tremor above the background buzzing of the wire.

Here, again, my memory is inconsistent. I can't say, for instance, whether Rae's comment brought back the earthquake, or if it was already present, emitting signals from some subterranean layer of my consciousness, like a psychological precursor, a cognitive electromagnetic pulse. What I do know is that I laughed and told her, yes, there had been a quake, but it was less than nothing, so small it hadn't woken up the kids. Although I must have thought this information would relieve her, it only seemed to agitate her more. "They didn't wake up?" she said. "How is that possible? CNN is reporting a 7.0 in Los Angeles."

"A 7.0?" I asked her. "Are you sure?" A 7.0 would be a massive earthquake, the fourth largest to hit Southern California in the twentieth century, a temblor three times as powerful as Northridge, nearly approaching Landers size. A 7.0 would have shaken books off shelves, shattered plates and glasses, stirred my children to a screaming cacophony of terror as their beds bounced and their toys began to dance like living creatures, like the substance of a nightmare come to life. A 7.0 would mean fallen freeways, pancaked buildings, cars crushed in parking structures, loss of property, loss of life. I thought back to the shaking, to its gentle rolling motion, the way it hadn't even pushed me out of bed. Was this another blurring of the line between memory and imagination? Had I somehow misread what was going on?

Phone in hand, I went to look in on my son and daughter. Both were still asleep, sprawled across their blankets as if this were any other day. Idly, I flicked on and off the hall light. Then, I told Rae that CNN must be wrong.

We talked for a minute longer—Rae and I each in our own way still uneasy—until I said that I was going back to bed. Yet after hanging up, I found myself drifting through the house with silent footsteps, like some strange ghost of earthquakes past. It reminded me of the first few hours after Northridge, when Rae had slept while I listened to the initial reports on a Walkman, watching the night-black sky yield piece by piece to the creeping glow of dawn. What I recall most about those moments was the quality of the darkness, the way that, with all the power out, there were no lamps, no noise, not even the glimmer of a human presence, nothing to penetrate the inky thickness of the night. Standing at my bedroom window, I had imagined that all of Los Angeles might be gone, erased, every structure fractured, an entire cityscape collapsed. When the sun rose, I found myself amazed at how intact L.A. appeared in the daylight, how undiminished, how . . . *normal*, so unlike what I'd supposed.

Today, however, it was different: whereas with Northridge, I had recognized the severity of the shock and been surprised by the lack of damage, here it appeared to be the other way around. I looked out the living room window, but the street offered no clues beyond the usual early morning silence, sun inching up behind the palm trees, porch lights on, a bike left in the driveway next door. The *Los Angeles Times* had been delivered, but the front page bore no reports of temblors, large or small. Finally, I turned on the television, where, over an aerial shot of an extensive ground rupture (twenty-five miles long, I would learn later) and some footage of a derailed Amtrak train, a male news reader reported that the quake had struck the Mojave Desert, between

Twentynine Palms and Barstow, 120 miles east of Los Angeles. The Mojave's not L.A., I remember thinking, and for one long second, I felt a flash of satisfaction, as if my perceptions had been confirmed. But before the sensation had a chance to settle, the broadcast cut to a map of the affected area, as the commentator recapped the available information about the earthquake, beginning with the fact that it was, indeed, the fourth largest Southern California temblor of the twentieth century, coming in at a magnitude of 7.0.

I was not the only person caught off guard by the Hector Mine earthquake. This seems something of an obvious statement, for that's what earthquakes do: they surprise us, shake us up, cast all of our illusions into doubt. Yet even if we start with that, Hector Mine is a little different, for what confounds us is less the quake itself—its physical or psychological effects, its devastation—than its unobtrusiveness, its location, its evolution, even the fact that it took place at all. If you read newspaper accounts of the temblor, the first theme that emerges is how calm it was, how easy, as opposed to, say, Northridge, which announced itself with a series of violent up-and-down jolts, as if the world were being kicked from underneath. "It wasn't like the Landers earthquake in 1992," David Harper of Landers told the *Los Angeles Times* on Sunday, October 17. "That one was sharp, like a truck ran into your house. This one felt like a ship on the ocean. Nothing broke. It was real mellow for a major earthquake." Then, as now, these comments offer me a measure of relief, of confirmation; if this man, living less than thirty miles from the epicenter, could misconstrue the severity of the tremor, it's little wonder that I did, as well. Yet such a situation also makes me wonder about our capacity to talk about earthquakes, to develop any kind of common vernacular. How can an interpretive scale like magnitude mean anything if a 7.0 is less extreme than a 6.7, even at the very flash point of the quake? How can we reflect the physical reality of an earthquake, the strength of the shaking, the temper of the slip? On the most basic level, this further illustrates the essential intractability of the experience, our inability to address the geologic in human terms. At the same time, it also highlights one more time the nebulous area of influence, the way a whole host of factors can contribute to our perception of an earthquake, and the force with which it is or isn't felt.

This idea of influence, of association, takes us in all sorts of odd directions when it comes to Hector Mine, amplifying the unexpected nature of the quake. Here, it seems, every action has an equal and opposing reaction, while every detail leads to some potential dichotomy. This is an earthquake, after all, that took place on a fault, the Lavic Lake, long considered to be dormant, a system that, until 1999, had remained quiet for longer than human civilization existed on the earth. But this is also an earthquake that was predicted two months before it happened, in an essay written by Cloud Man, and posted on his site. These two facts are utterly discordant, and yet they coexist in the subtext of the event. I'm not sure what that suggests, except perhaps the intervention of chaos, complexity, connections just beyond the range of human understanding—in other words, geopoetry. Adding still another layer to the enigma is the fact that Hector Mine was epicentered at the U.S. Marine Base at Twentynine Palms, where marines were in the middle of three weeks of live-fire combat training, which was disrupted by the quake. Every time I think of that, it brings to mind a bit of earthquake lore Jim Berkland once shared with me, about temblors being triggered in Colorado during the early 1960s as a result of nuclear tests. I've never been able to verify Berkland's story, but I'd be lying if I said it didn't resonate, if only because of all the ordnance near

Hector Mine. When geologists from the USGS and NASA's Jet Propulsion Laboratory examined the surface traces, they were accompanied by two Marine Corps weapons experts "to ensure that when they landed periodically to scramble on foot along the new rupture, they did not step on mines or unexploded shells." Were I more conspiratorial in nature, I might be inclined to read between the lines of such a statement; if oil drilling and water extraction can induce seismicity, why not the detonation of explosives in a seismic zone?

Of course, the problem with this brand of conjecture is that, while it may be fun to play with, it inevitably leads us into fantasy, rather than legitimate research. A more useful, if no less unanticipated, area of speculation involves the relationship between Hector Mine and Landers, which spans a distance of seven years and twenty-five miles. As early as two days after the 1999 earthquake, the *Los Angeles Times* had already begun reporting on the possibility of a connection, citing researchers like Ross Stein. On October 18, the paper even reproduced one of Stein's electronic simulations beneath the heading "Too Much Stress"—a multicolored stress map of the Yucca Valley, with stars marking the Big Bear, Joshua Tree, Landers, and Hector Mine temblors, and bubbles of red and purple delineating the distribution of the strain. In the upper right-hand corner, the Hector Mine epicenter sits in a side lobe that balloons out from the Landers rupture, as clear a model of seismic interaction as you might ever see. "This triggering from one earthquake to the next," Stein told *Times* science writers Robert Lee Hotz, Diana Marcum, and Kenneth Reich, "is important in understanding earthquake behavior. Seven years after Landers, another quake pops off and it pops off in a region brought closer to failure by the Landers quake." In the link between Landers and Hector Mine, interaction appeared to have come full circle, since it was only in the wake of the initial earthquake that most seismologists had given credence to the theory at all.

There's a lovely symmetry to that dynamic, a hint of resolution you don't often see with earthquakes or geology. It's like tracking a line you expect to be discontinuous, and discovering that it proceeds, unbroken, from points *A* to *Z*. Still, for all the satisfaction this may offer, the closer we look, the more we find ourselves in another gray area, confronting a whole new set of difficulties. Despite the apparent evidence of interaction, Hector Mine is an anomalous earthquake, one that probably shouldn't have happened—at least not as it did. First, there's the issue of time, which is inconsistent on a couple of levels, beginning with that interval of seven years. A seven-year gap is hardly unheard of among related earthquakes; all we need to do is look at the seismic history of Turkey or the San Francisco Bay Area. In the case of Landers, though, interaction (or what we've seen of it) has unfolded at a somewhat faster pace. "The interesting thing," notes Ross Stein, "is that after Landers, triggered earthquakes occurred immediately, within a matter of days. The big one didn't pop off for seven years, although it's probably part of the process. But that's disputed by all kinds of people. There's a lot of debate." Even if we accept the model of interaction, we still have to deal with the question of proportion, of how so much strain could build up on the Lavic Lake so quickly after Landers, when the fault had been essentially silent through countless earthquakes over ten millennia. "Here we have a minor fault producing a major quake, which is disturbing," Stein said shortly after the rupture, a sentiment that continues to reverberate as seismologists attempt to puzzle out what forces are involved here, and integrate them in a cohesive vision of events. Hector Mine, in other words, might have been influenced by interaction, but there are questions as to whether interaction can account for the earthquake *in and of itself.*

The simplest solution is to interpret Hector Mine as a distant aftershock of Landers, an approach that, Stein and others have suggested, may be less radical than it sounds. Aftershocks, after all, can occur years, decades, even centuries beyond a mainshock; there are still temblors in the Midwest, for instance, that seismologists believe to be aftershocks of the New Madrid sequence, echoing through the lithosphere like whispers feeding back on themselves. "One extreme of that idea," says David Bowman, a geophysicist at California State University, Fullerton, "is that background earthquakes, background seismicity, are really just aftershocks from very old earthquakes. And that everything is an aftershock from something else." Yet the trouble with framing Hector Mine as an aftershock, Bowman suggests, is that it's a desperate measure, an attempt to fit the data to a preexisting thesis, not unlike the miscalculation Allan Lindh made in regard to the 1934 Parkfield quake. If anyone understands this impulse, it would be Bowman, for he was, out of necessity, an early advocate of the aftershock theory; Hector Mine struck three days before he was to defend his PhD dissertation, in which he intended to model an entirely new form of interaction, one based on the acceleration of stress throughout a region leading up to a quake. "I remember thinking, what a great pièce de résistance for my thesis," Bowman tells me somewhat ruefully one afternoon in Fullerton. "Then I ran the numbers, and they didn't fit. You want to see somebody sweating bullets? That was very difficult for me. And the way I waved it off was to say, well, Hector Mine and Landers are essentially the same event because Hector Mine is a gigantic aftershock. But I was kind of forced into that argument, which is an uncomfortable position to be in."

Bowman tells me all this from behind a desk in his university office, and as he speaks, he carves elaborate shapes in the air with his hands. His eyes glitter behind round John Lennon–style glasses; every now and then, he punctuates his comments with a loud, snuffling guffaw. Although he's in his early thirties, he looks much younger, his features smooth, not quite hardened, as if he's still carrying a layer of baby fat. His demeanor, too, is less professorial than like a precocious graduate student, words erupting in a long, low rumble, propagating one to the next with an edge of urgency, excitement, as if were he to speak more slowly, the ideas that seem to swirl around him like bees buzzing might simply fly away.

In many ways, Bowman's exuberance is only fitting, for he is at the very start of his career, new to this appointment, fresh off a year and a half of postdoctoral work at the Institut de Physique du Globe in Paris, where he went after his successful thesis defense. His office has the aura of arrival; small, windowless, tucked into a third floor corner of McCarthy Hall, it is neat, nondescript, and virtually empty, as if Bowman had not yet settled in. On a shelf behind his desk sits a row of books about seismicity, including one thick volume called *Earthquake Prediction*, and a copy of Christopher Scholz's *The Mechanics of Earthquakes and Faulting*, which I've been trying to read for months now, although its mix of technical jargon and endless mathematical formulas has proved as daunting and ineluctable as earthquakes themselves. The walls are unadorned except for a single poster featuring a satellite photograph of Venus, the surface of which is latticed with an intricate network of thin black scars. "What are those?" I ask, as I stare up at the photo, trying to identify a logic in the lines. Despite my efforts, the image remains resistant, like one of those hidden picture drawings that looks like little more than a scribble, indecipherable, chaotic, until you unfocus your eyes and the shape of a car or a clown or a building emerges from the background noise.

"Those are actually faults on Venus," Bowman replies, eyes popping just a little, mouth curling into a narrow grin.

"Really?" I say. "There are faults on Venus?" I've never heard this, but Jim Berkland has told me about moonquakes, so I'm not entirely surprised.

"Yeah," Bowman continues, and his voice grows high and reedy, as if he can't answer my question fast enough. "Venus and Earth are in many aspects twins. They're the same mass, they're the same composition, they're in the same sort of neighborhood in the solar system, and these things really matter for a planet. Earth and Venus are much more alike than, say, Earth and Mars."

"So Venus has fault structures similar to the earth's?"

"Well, see, this is what's really interesting. Because Venus is very different in its geologic evolution from the earth. On Venus, you don't have a lot of active fault structures that transfer stress horizontally. Most of the faults are locked into place. There are no strike-slip faults, no thrust faults, no activity going on at all. In fact, right now, as far as we can tell, the planet is essentially dead."

"Dead?" Beneath the surface of my consciousness, I feel an idea begin to nucleate, to generate friction, although I can't quite put my finger on what it is. "You mean the way Lavic Lake was supposed to be?"

"Maybe," Bowman says, and chuckles just a little. "There are any number of different possibilities."

When Bowman mentions possibilities, what he's really talking about is a set of windows opening onto the inner life of earthquakes. For him, the idea of faults on Venus is not a corollary to such a process, but the place it all begins. Bowman, after all, came to geophysics from the planetary sciences; as a kid, he was infatuated with the cosmos, with Apollo missions and Carl Sagan on TV. It wasn't until he was an undergraduate at the University of Southern California that he took his first geology class, and even then, the attraction was less the earth *per se* than the idea of studying the way planets—all planets, any planets—work. "I got suckered into geology," he says. "At USC, I took Charlie Sammis's geophysics class, and when I was looking for a senior thesis, Charlie took me aside and said, 'Why don't you leave astronomy, that silly dead science, and do something interesting?' " Bowman laughs at the memory, but in his own way, Sammis was absolutely serious; an iconoclastic figure, he considers seismology a dynamic science, in which we have no choice but to push the boundaries, to follow hunches, to look for connections in the least expected places, and then apply them to the movements of the earth. In early 1995, he became the first mainstream seismologist since James Whitcomb to issue a specific earthquake forecast, calling for a temblor in the 6.0–6.5 range to strike the San Andreas in Central California, perhaps in the area of Parkfield, during a 160-day period between February 1 and July 9. Although the prediction didn't pan out, Sammis remained unapologetic, arguing that his research had been valid and the danger real. As for Bowman, it was Sammis who suggested that he work on Venus, both to understand the faults there in their own right, and to see if they might lead into a larger view. "We were looking for a planetary physics," Bowman tells me—which means a physics that might also fit this world.

This concept of physics—a physics of seismology, of earthquakes—is perhaps the central aspect of Bowman's research, and, in large measure, what sets his work apart. "All of tectonics," he says, "all of faults, volcanoes, everything we see on the surface of the planet, is just the planet's effort to lose heat. Because there is heat on the inside of the planet, and it wants to get out. Plate tectonics are a byproduct of that heat recirculation. And earthquakes are a byproduct of plate tectonics. So essentially, the engine that's driving everything is this effort to cool the planet off. And most geologists don't think of things that way." Once again, the issue here is practicality—or more accurately, how to balance practicality with a theoretical edge. When we look at earthquakes, after all, it's virtually impossible to strip away the human component, to separate ourselves from what has happened, from what is going to happen, to view these periodic burps and spasms as parts of processes that have *nothing to do with us*. This, it turns out, is what Venus offered Bowman, a theoretical laboratory in which to push his thinking beyond the more pragmatic concerns that can't help but assert themselves when we investigate the earth. The thing about seismicity, however, is that it always surprises us, that it is capricious, serendipitous, taking us in directions we never meant to pursue. I can't help considering this as I ponder Bowman's passage through the seismic landscape, which, in much the same manner as Ross Stein stumbling upon interaction, has evolved out of its own peculiar web of influences and implications, beginning with Northridge, which interrupted the Venus work in January 1994. "All of a sudden," Bowman remembers, "I found myself out in the field, rubbing elbows with a lot of earthquake people, and it really started to interest me. Then in 1995, Charlie Sammis had this great idea on earthquakes, on how they interact in preparation for a large event. It was even more exciting. He was working with a statistical physicist, bringing back ideas I had abandoned as an undergraduate, and he asked me to work on it. At first, I only got involved part-time because I still wanted to go into planetary physics. But after a while, I got sucked in."

The idea that Sammis had—and that Bowman eventually ran

with—is a bit of theoretical thinking so profound and yet so simple it's astonishing no one ever came up with it before. At its heart is the premise that, during the years before a major earthquake, stresses in a region align themselves, as if to prepare for the coming break. On the surface, this doesn't sound all that different from Ross Stein's theory of interaction, with its notion of stress transfer in the interplay of earthquakes and related faults. Yet while Stein's focus is primarily on significant earthquakes (which are responsible for most of the energy released in a seismic landscape), Bowman and Sammis were interested in what is known as intermediate seismicity—lesser temblors, generally in the 3.5–5.5 range, which are more frequent and easier to track. "If you look at a place where there's going to be a large earthquake," Bowman explains, "there are other, smaller earthquakes that happen in the background. Let's take the Landers earthquake, as an example. Landers was out in the Mojave Desert, where there are a lot of earthquakes. The question is, if you take a region of hundreds of kilometers around the Landers epicenter, can you see any difference in the seismicity, in the earthquakes, before the shock? We're looking for changes in the rate, the numbers, the size of background earthquakes over a period of tens of years."

At its most elemental, what Bowman is describing is an inside-out view of interaction, one involving less stress transfer than stress buildup, or, in his phrase, "accelerated motion release." This is a small distinction, but a key one, for it suggests that rather than merely identifying interaction after the fact, we may be able to see such relationships as they develop—to push interaction into the future, as it were. Of course, given the elliptical nature of earthquakes, it seems only appropriate that, in order to pursue this, we would first have to turn to the past. That's precisely how Bowman and Sammis tested their hypothesis, by examining eight California earthquakes, beginning with the 7.5 Kern County quake of July 21, 1952, and continuing up through Loma Prieta, Landers, and Northridge. As a safeguard, they also looked at "a small set of earthquakes outside the space-time window of our study," including the 1906 San Francisco temblor, as well as a thousand computer-generated earthquake catalogues, each consisting of one hundred simulated events, a process meant to establish the statistical validity of their model. What they discovered, in both the real world and the virtual, was nothing short of incredible: in every case, there was a pronounced acceleration in regional seismicity prior to the earthquake, with intermediate earthquakes increasing in volume and frequency throughout the years leading up to a main event. As Bowman and Sammis conclude in a 1998 paper cowritten for the *Journal of Geophysical Research* with three French geophysicists:

> While the acceleration pattern found for each single earthquake could be the result of chance with a probability close to 1/2, when taking all together, we find that our results reject the null hypothesis that the seismic acceleration is due to chance with a confidence level better than 99.5%.

Such a statement brings us back to the realm of physics, theoretical physics, a physics so theoretical that, initially, it included almost no geology. In fact, Bowman explains with a slightly sheepish cachinnation, the original model for his research was not at all seismologic, but rather "the statistical physics of a critical phase transition." In this context, he writes, "the word 'critical' describes a system at the boundary between order and disorder and is characterized by both extreme susceptibility to external factors and strong correlation between different parts of the system." Perhaps the most obvious example of that in the natural world is the process by which water freezes; as the temperature gets colder, water crystals develop as ice on the microscopic level, without ever having had to reach the freezing point. This is a critical phase change, the molecular shift from liquid to solid, in which the very essence of a substance is transformed. The same, Bowman suggests, appears to be true of earthquakes, at least in the way stress accumulates until it hits the moment of critical release. "It's a very loosey-goosey analogy," he admits. "But if you do the blurry details of critical point theory, you find that there's an acceleration in activity before any phase change, whatever it may be. Accelerating seismicity is an increase like that, and if you take the energy released in earthquakes, precursory earthquakes, throughout these huge regions before a large one, you should see an acceleration in the energy release."

Needless to say, removing geology from earthquake analysis in favor of statistical physics was, as Bowman suggests, "a pretty radical act." In his earliest models, he even used statistics to determine the size of seismic regions, eschewing physical data in favor of a computer logarithm that essentially grew a series of concentric circles around the epicenter of a particular temblor, then gauged seismicity based on the number of intermediate earthquakes that fell within those boundary lines. "The idea," he says, "is that, at every circle, we plot seismicity. If it's accelerating, that's a plus. If not, that's a minus. Then we quantify the whole thing by calculating how much it's accelerating. And just before each earthquake, we let the computer pick out the best radius, where we see the most acceleration." Although Bowman has defended the strategy as a way of "letting earthquakes define their own regions of seismicity," some in the seismological community warned that it was just a heightened form of pattern recognition, the Texas sharpshooter problem all over again. This is the same issue faced by Allan Lindh at Parkfield and Ross Stein in his work on the North Anatolian, as well as, in many ways, by predictors like Jim Berkland or Cloud Man. How do we know what we are seeing? How do we tell the difference between order and illusion—especially when you consider the paucity of seismic history, the inconsistency of monitoring networks, our inability to gain any accurate purchase on the past? These questions, Bowman admits, are impossible to get away from, precisely because people and machines are so adept at interpreting patterns, including many that are spurious, like the belief that big Southern California earthquakes happen only in the morning, a myth fueled by the timing of the area's last five large-scale tremors: Whittier Narrows (7:42 a.m.), Sierra Madre (7:43 a.m.), Landers (4:58 a.m.), Northridge (4:31 a.m.), and Hector Mine (2:46 a.m.). Nonetheless, two important details helped to validate Bowman's research—first, the statistical unlikelihood of so many earthquakes randomly reflecting accelerating seismicity, and second, that the magnitude of each event was directly proportional to the size of its region, or, as Bowman tells me with a slight flourish, that they "scaled."

The significance of this "scaling" cannot be overstated; it's the element that brings everything into focus, that carries what Bowman calls "the smell of truth." What it suggests is the presence of some underlying physical process, a kind of symmetry or balance, a method to the workings of the earth. Even more, it appears to mirror other observations about earthquakes, from the Gutenberg-Richter relation and Omori's law of aftershock decay (a theory first posited by Fusakichi Omori following the 1891 Nobi, Japan, earthquake, which demonstrates that the rate of aftershocks falls off exponentially over time) to the findings of British geophysicist Geoffrey King, who, in the 1980s, applied the principles of fractal geometry to seismicity and discovered a proportional relationship not only between the size of earthquakes, but between the size of faults as well. You can read these parallels any way you want to, in either practical or philosophical terms. Bowman himself would eventually adapt the Gutenberg-Richter relation to his own purposes, arguing that its sense of seismic distribution—the way there are ten times as many 3.0s as 4.0s, and ten times as many 4.0s as 5.0s—offered a logic for how accelerating seismicity pushed a region to the point of failure, since "having more and more earthquakes also means you have a higher probability of large earthquakes." To me, though, it's like looking into the face of mathematics, or maybe it's the face of God. In Jewish mysticism, God is often referred to as a verb, as existing at the intersection of act and language, the word that literally invents the world. That's an idea I've always loved, for the power it bestows upon imagination, its inherent faith that if you can say a thing, you can make it so. Still, when I think about the clarity of scaling, I can't help but believe that God is more than just a word. Verb or not, I tell myself, God must be a number, too.

Yet number or no number, God is a trickster; God, in other words, throws dice. "If you want to make God laugh, tell Him your plans," the adage goes, and if ever such a statement had a real-world analogue, it would be Bowman's experience with Hector Mine. "According to the statistical physics model we'd developed," Bowman says, "Hector Mine didn't have a preparation phase. When we plotted the circles around it on the computer, there was no acceleration. The whole region was dominated by a Landers aftershock, which had actually caused stress to decelerate." This is the quandary Bowman faced as he prepared to defend his dissertation, a quandary made all the more difficult by the theoretical nature of his investigations, which stand at what he calls "the bleeding edge." Here, as with Ross Stein's study of interaction or Kerry Sieh's paleoseismological work, we see up close the vagaries of earthquake research, the interplay of fact and nuance, of observation and intuition; this is science as it happens, a matter of trial and error and trial again. For Bowman, the breakthrough wouldn't come until he got to Paris, where he began what has become an ongoing collaboration with Geoffrey King. Under King's tutelage, he undertook a comprehensive testing of his theory, looking at earthquake catalogues from across geography and history, hunting for an explanation that might resolve the contradiction of Hector Mine. "Initially," Bowman explains, "we went through all the earthquakes that we knew of, to see if they took place in regions of increased stress. The problem, though, is that you can never have a good enough historic record. No matter how well you interpret the data, if you make one small change, it completely throws the stress patterns, and your answers, out of whack. The small errors make a big difference in this, which is, of course, a sign of chaos. But basically, what it means is that when you run the model forward, it doesn't work."

The turning point, as it happened, emerged from yet another apparent bit of desperation—the decision to go back to geology, to adjust the model to blend statistical and observational science, to bring seismicity back into the mix. To be fair, Bowman seems to have had an inkling of this from the earliest stages of his research: "The present work," he and Sammis wrote in 1998,

> can be improved in many ways. First, there is no physical reason for the critical region to be a perfect circle centered on the epicenter of the final event. We have already extended this analysis to use elliptical regions rather than circles, with no significant modification of the results. A better approach would be to use the natural clustering of the seismicity to define the regions. . . . These fluctuations permit a more precise

> characterization of the precursory seismicity, which would consequently enable a more accurate determination of the critical region.

In Paris, however, he and King took that idea and pushed it, not only using seismicity to determine regions, but developing a strategy to analyze pre-earthquake stress levels by essentially turning back the seismic clock. "The usual way of doing things," Bowman tells me, "is to calculate the stress field after the earthquake. That's how Ross Stein or Ruth Harris maps out interaction. They look at the stress created by a particular earthquake, then try to see how subsequent seismicity relates. What Geoff and I said was, 'Let's do it backwards.' Because mathematically, it's the same thing. If you take an earthquake and slip it the opposite direction, back to the starting point, what you get is the stress field that must have existed before the earthquake took place."

"So you can simulate that on the computer?" I ask. "Basically undo the earthquake?"

"Undo the earthquake."

"And what you're left with are the circumstances in which it happened?"

"Exactly," Bowman says. "That's the assertion we're making. If you run the earthquake backwards, you can get the stress field, and then you can identify regions of high stress before the earthquake. That's the thing we never did before. It's the piece of the puzzle we never had."

Bowman beams at me from behind his desk, expectant, like a young boy with a new plaything. As I smile back, I think about the movie *Superman*, in which the Man of Steel sets the earth spinning backwards on its axis to undo a major temblor on the San Andreas that's been triggered by a Lex Luthor–created nuclear bomb. The image evokes so many resonances I don't know where to begin: with the bomb, which suggests both Hector Mine and Colorado (all that military ordnance) or the idea that time can be so malleable, so fluid, that it can be read back and forth, in so many complex ways. In the film, Superman has selfish motivations; he's not out to save humanity, nor even California, but rather Lois Lane, who has crashed her car into the fault at the moment of the rupture, only to be crushed to death between the grinding plates. By reversing the earth's rotation, he effectively turns back time to the instant right before the earthquake, at which point he swoops in from deep space to save the day. This, of course, is a ridiculous premise—even if Superman could invert the revolution of the planet, how would such an act rewind chronology when the rest of the space-time continuum goes on undisturbed? Yet while that may make for a problematic piece of science fiction, it does provide its own unanticipated filter, a three-dimensional symbol of the work that Bowman and King are doing just the same.

After all, to listen to Bowman talk about undoing earthquakes is to interact with eternity all over again, albeit on a brand-new level, one that casts the infinite in starkly finite terms. It is to mitigate the inscrutability, to gain some perspective, to exist not at time's mercy but in its measure, relying less on history than on seismicity, which is especially useful in regard to the Mojave, where history goes back only so far. Again, this is a subtle distinction, but a profound one, for it effectively alters our frame of reference, making the whole issue of earthquakes appear less open-ended, more resolved. "If you have the stress distributions beforehand," Bowman explains, "then you can identify areas that are high stress, positive, and also those that are negative, low stress. What we did was to look at all the regions of positive stress changes—regions, in other words, that were high stress before their earthquakes—and then check the seismicity, the actual physical precursory activity, to see how it matched up. We redid all the tests we had previously done for California, and they worked." Even more significant, the act of rewinding earthquakes is what we might call event specific, which allows Bowman to narrow his focus to a particular temblor and effectively isolate it from the background noise. "The problem with circles," he says, "is that you're listening in areas where you shouldn't be, so you're contaminated by other stuff. But when you put geology back into the problem, voilà . . . you get a much more accurate picture of acceleration than circles provide." The perfect example of this is Hector Mine, which now works because it sits in a region defined by its own seismicity, rendering it fundamentally distinct from the Landers quake. "The fact of the matter," Bowman concludes, "is that Hector Mine may have been mainly stressed out from some earthquake two hundred years ago, five hundred years ago, that we don't know about. It could have been ready to go all along."

As Bowman talks, he reaches to the floor beside him and withdraws a laptop from a shoulder bag. I watch him fold the machine open and boot it up, listen to the telltale tones and whirs. For a moment, I'm reminded of Ross Stein and his simulations, of sitting in his Menlo Park office while he clicked back and forth through earthquake history. There's a certain irony to this, or perhaps it's just another metaphor, a sign that, even with geology back in the equation, we still have to interpret seismicity at a tangent, to fill the gaps in our comprehension with imagination or technology. "Let me show you something," Bowman says, moving a finger across the mouse pad, bringing up an empty field of green. "These are what the stress fields look like," he continues, and after I lean in close to get a better angle, he taps an arrow button and takes me through a brief series of images before coming back to the opening screen.

At first, I'm not sure what I'm seeing. Unlike Stein's computer sequences, these look more like fractals, like geometric abstractions in patterns of blue and red. "Here's the stress before the earthquake," Bowman declares, showing me an early iteration, which has the shape of a pair of wings. Or not wings, exactly, but something else, something I recognize, although I can't pinpoint what it is. As Bowman clicks us closer to the rupture, I watch the wings grow, their red, or stressed, lobes becoming bigger and more rounded, until the quake happens and the color field is reversed. Now, the blue lobes are the most pronounced part of the image, indicating the deceleration that follows a temblor, while the red fall off to little filaments, like small, thin darts of cybernetic flame. "If you make a map," Bowman says, pointing to the monitor, "a picture of the stress field, one possible solution is that it looks like a butterfly. And if you draw a circle around the butterfly, you'll get a lot of area that's not related, but the circle will be the same you'd get from the old way of doing it."

I look at the screen again as Bowman reruns the sequence, wondering how I could have missed it. Butterflies, of course—that's what they are, wings spread out as if in full flight. I can almost see their electronic bodies moving, stirring up the virtual air around them, causing unknown and unknowable variations in the way we think of earthquakes, like a macrocosm of the butterfly effect. Even their appearance feels like it means something, as if I have walked into a life-size Mandelbrot set. "Cool," I utter, almost entirely unaware of what I'm saying, drawing out the syllable like a starstruck kid.

Bowman laughs. He thinks I'm talking about the circles, the way they scale so precisely with the stress fields, like some final proof that the patterns have been inherent in the process all along. "That's exactly what we said," he remembers. "That was our 'Holy shit, we hit the jackpot' moment. Again, it gets your nose twitching. You know you're onto something here."

There's a fleeting instant—less than a second, really—when I almost interrupt him, when I feel compelled to tell him how our logic lines have crossed. But before I can act upon the impulse, I realize that it doesn't matter, or, more intuitively, that it all connects. What both Bowman and I are marveling at, after all, is yet one more unexpected appearance of order, even if, with the butterflies, it's an order that is largely invented, an imposition of my own subjective mind. In science, as in art or love or language, we tend to find what we are seeking; we are incapable, William S. Burroughs once suggested, of observing anything that we don't, on some cellular level, already know is there. This is not to say that Bowman's order is solipsistic—in fact, I'd argue, just the opposite is true. But the reason it comes as such a revelation is that most contemporary geology is a matter of small-picture thinking, in which investigators identify a narrow area of inquiry and "work it to death," as Ross Stein puts it, without ever raising their heads to trace a line of intersection to a broader landscape. "You can't look at earthquakes," Bowman says, "in isolation from each other. And you can't look at earthquakes in isolation from the faults on which they occur." Listening to him, I find myself thinking that the same is true of earthquake *research*, that you can't look at ideas in isolation either, that all these theories and cogitations have their own influences and interactions, that they, too, are part of the complexity of the seismic world.

In that sense, Bowman's work marks a place where seismology starts to come together, a unified field theory of earthquakes, as it were. It is, in other words, an attempt to present an *intellectual* as much as a *geologic* order, to gather the existing elements of seismic research and listen for their echoes, tracing what we might call an acceleration of ideas. Certainly, Bowman is explicit about the debt he owes Ross Stein and his theory of stress transfer and regional seismicity; remove that as a starting point, and the very notion of accelerating motion release might not exist. Yet equally implicit in his thinking are cornerstones like Harry Fielding Reid's elastic rebound theory—without which even the *concept* of interaction or acceleration would lack a necessary context—or Kerry Sieh's excavations of the San Andreas, which definitively established that temblors occur cyclically, that the sequence of tension and release first postulated by Reid after the 1906 earthquake unfolds according to a pattern of some kind. Then, there are all those endless foreshocks, thousands of years of them, described in anecdote and scientific documents, many of which, Bowman believes, represent evidence of accelerating seismicity—or would have, had there been any systematic way to measure strain buildup during the months and years prior to the mainshocks they forebore. This, I think, is a particularly compelling observation, one that offers a strategy for rethinking an event like the Haicheng earthquake, while amplifying the irregularities Allan Lindh noted before Loma Prieta, framing the Lake Elsman tremors as, among other things, textbook earthquakes of intermediate magnitude. For Bowman, however, these ideas are incomplete without one another, and if we mean to understand the way earthquakes work within a geographic area, we need to look at interaction in the broadest possible terms. "Stress transfer," he says, to cite one example, "is only half of the problem. Stress loading is the other half. Transfer is too small a lens. It's seeing earthquakes as triggering other earthquakes, when we should also be looking at how the loading rate influences background seismicity throughout a region, slowly amping everything up."

Still, even here, amid all this confluence, we can't help but find ourselves thrust up against the contradictions of seismicity, if only because there's so much we don't know. Partly, this has to do with chaos, which—as is only to be expected—becomes increasingly a factor the more closely we look at how acceleration ricochets beneath the earth. More important are the vagaries of chronology, the dichotomy between human and geologic time. Seismicity, after all, unfolds across aeons, developing at a pace so far beyond us that it often doesn't seem to be developing at all. "The way I'd describe it," Bowman says, in an echo of Stein or Lindh or even Lucy Jones, "is that, on any fault, there is a time when it is ready to go, when it has sat around, accumulating strain, long enough. But when it goes precisely depends on how it interacts with other faults. Kerry Sieh's work is a case study in that. You have a 135-year recurrence interval on the San Andreas, but there's a scatter of fifty years on either side that has to do with interactions with other events." Adding a further layer of complication is the fact that the entire system, from plates to faults to stress accumulation, is always shifting, always moving, which guarantees that no two earthquake cycles, even those that strike the same fault segment or seismic region, are ever fully alike. As Bowman puts it, "Seismicity is not static; it's dynamic. The patterns are always evolving, and they evolve in geologic time. This evolution in geologic time affects the way earthquakes happen, but the way earthquakes happen affects the geologic evolution, so there is this tension, this back and forth, between the two. All these planetary forces, planetary physics, unfolding across a million-year timescale—that's the way seismicity works."

As if to illustrate the point, Bowman waves his left hand at the poster of Venus, which I've forgotten until now. "Let me give you an example," he says. "You remember when I told you Venus was a dead planet?"

"Yes."

"Well, the truth is . . ." He flashes a quick grin and looks a little sheepish. "Let's just say it might be dormant instead."

"Dormant?" I ask, not understanding. "How can something be dead and dormant all at once?"

"That's just it," Bowman answers. "If you look at a ten- or a thousand- or even a million-year timescale, Venus doesn't behave like a living planet. But it may operate on a more extended cycle than we can wrap our minds around." What makes Venus so hard to fathom, he goes on, is that its entire surface is, for all intents and purposes, of a uniform age, "something like 300 million years old." This is in stark contrast to Earth, where, because of seismicity and plate tectonics, rocks from almost every era of the planet's history are constantly churning up and down. "The oldest known continental rock," writes John McPhee in *Annals of the Former World*, "was discovered east of Great Bear Lake, in the Canadian Northwest Territories, in 1989, and has a uranium-lead age of 3.96 billion years. The earth itself, according to radiometrics, is six hundred million years older than that." If, as Bowman has suggested, Earth and Venus are similar in so many other respects, how could they be so different when it comes to this?

"Maybe," I venture, "Venus died 300 million years ago, and what we're looking at is just the residue of that event."

"That's one possibility," Bowman says. "The other possibility is that Venus, in contrast to the earth, sits really nice and stable for a few hundred million years, and then all of a sudden burps. The whole planet just overturns."

I try to imagine what this would look like. I try to imagine the surface of Venus exploding, going from wholly stable to wholly unstable in a single geologic breath. I try to imagine a timescale so broad it works in 300- or 400-million-year cycles, each of which is one-fifteenth the age of the planet itself. I try to imagine all these things, but I can't do it, can't push my mind from this room, this life, this human time frame, or maybe it's just that I can't see my way clear of Earth. Eternity, that's what we're talking about here, eternity on the cosmic level, but unlike the eternity I've glimpsed inside the San Andreas or looking at Stein's and Bowman's seismic models, this one is too abstract, too supernal, for me. For a moment, I'm reminded of Jim Berkland and his Machu Picchu meditations, with their hint of revelation, of some secret handed down from up above. In the end, I suppose, I prefer my eternity earthbound; I'm driven to look for God not in the sky, but in the ground.

Yet once more, just when I think I've come to some conclusion about earthquakes, I have to rethink my terms. Venus, after all, *is* the ground, or so Bowman reminds me—a planet just like Earth, subject to the laws of planetary physics, which, regardless of how variously they may manifest, remain consistent throughout the universe. In that sense, it is less a twin of Earth than a celestial funhouse mirror, in which our world reveals itself by indirection, not through what it has in common with Venus, but rather what it does not. This is the value of studying it, 300-million-year cycle or no 300-million-year cycle, to observe the workings of another system, and, in so doing, gain some insight on our own. "What are the fundamental differences between Earth and Venus?" Bowman asks, taking a rhetorical posture. "For one thing, Earth has water and a cool surface. Venus has no water and a really hot dense atmosphere. And that affects tectonics, because water is one of the most important things allowing earthquakes to occur."

At the mention of the word *water*, something starts to shift again inside me, another bit of intellectual interaction, an idea. It's like a tickle, or a flutter in the pit of my stomach, the kind of sensation you feel in the face of an occurrence you can't quite believe. What Bowman has just done, after all, is to drop a plumb line straight through thirty-two hundred years of earthquake history, incorporating the full range of anecdote and observation from ancient Greece and China to the Haicheng earthquake, with the work of Paul Silver, Evelyn Roeloffs, and even Cloud Man scattered in between. Water, Bowman points out, functions as a catalyst for subcritical crack growth, chemically interacting with rocks before a temblor, getting inside small, subterranean cracks and causing them to grow. Water also acts as something of a seismic lubricant—"like oil," Bowman tells me—priming the dry, locked edges of a fault segment, which, in turn, makes it far more likely to slip. This is fluid-induced seismicity on a whole new level, in which, all of a sudden, water is no longer a precursor, an "indirect indicator," as Silver has called it, but a condition, a key piece of the mechanism that drives the engine of the earth. "If the earth were dry," Bowman explains, "truly dry in the sense that Venus is, plate tectonics as we know it would probably come to a stop. On Venus, it's hotter, plus there's no water, which is why it's locked into place. There's no water because it's so hot. And it's so hot because there's no water. It's a classic feedback cycle. On Earth, we're just right. We're in the perfect position in the solar system, the perfect distance from the sun, for water to exist as water, which allows plate tectonics to operate."

If you take such a comment on its surface, it can seem disturbing, inconsistent even: how could plate tectonics be a matter of perfection when it causes disruption, disorder, *chaos*, of both the scientific and the human kind? But the ideas Bowman is invoking operate on many different levels, much like seismicity itself. For me, what he's talking about are the dynamics of a living planet, the complex geometry of relationships—seismologic, geologic, environmental, and, yes, even chaotic—that contribute to the way the world works, to its constant malleability, its ebb and flow. A living planet is not fixed, but fluid. A living planet is always on the move. In some sense, you might even say that a living planet exists in a continual state of crisis, that this urgency, this insistence, is what defines it as alive. "The real question," Ross Stein has said, "is not why big earthquakes happen so frequently, but why they happen so *in*frequently. They should probably happen all the time." That, to be sure, remains open to interpretation, even in the context of stress loading and transfer. "We still don't know what causes an actual earthquake," admits Charles Sammis, "whether seismicity arises solely from strain buildup, or if the crust sits constantly near the breaking point." Still, if all this leads us to yet another place where our imaginations run up against the limits of our knowledge, I can't help being struck by an unexpected convergence, the way the circumstances necessary for life (earth, water, our exact location in the solar system) are also necessary for seismicity, as if without one, the other could not exist. Earthquakes, in other words, may disturb us, they may frighten us, shake us up, or even kill us, but on the most fundamental level, we might not be able to survive without them, for they are an expression of a living planet at its most profound.

In many ways, this is the geopoetic leap to end all geopoetic leaps, a leap not only of the mind, but also of the soul. At the same time, it helps explain a lot—even, to some extent, a quake like Hector Mine. "If you study paleoseismological records," Charles Sammis says, "it appears that large earthquakes cluster in time. You see it in the Eastern Mojave, and also in the Imperial Valley. What this could mean is that perhaps the entire region goes critical, and requires more than one earthquake to diffuse the stress. If that's the case, Landers and Hector Mine might work together, like a one-two punch." Such observations mirror those of Christopher Scholz and Ross Stein in regard to Southern California and the slow but steady rise in seismicity since the 1987 Whittier Narrows temblor; they also bring to mind the history of the Bay Area, with its increased activity in the seventy years leading up to 1906. If we look at them from the proper angle, however, they may take us one step further, towards a recognition not just of what is but what might be. Sammis, for one, believes that Bowman's research into regional seismicity could ultimately lead to what he calls "statistical forecasting," in which seismologists use the physics of stress loading and stress transfer to determine if a big quake is more or less likely in a certain location—although these "predictions," such as they are, would probably be framed in increments of years. Already, John Rundle and Kristy Tiampo of the University of Colorado at Boulder have developed a related model that has yielded four successful forecasts of tremors greater than 5.0—including a 5.7 that struck the Baja region of Mexico on February 22, 2002—but they, too, are working with an expanded time frame: their "seismic window" extends from January 1, 2000, through December 31, 2010. Bowman, meanwhile, has a student creating virtual temblors on a computer to calculate the stresses building up on literally every California fault. The hope, he says, is that it might offer a clue, a hint, some indication of where the next significant earthquake will be.

It's tempting to see this as some kind of intellectual turning point, a shift in the way we think about seismology, about the information that it offers, the assurances it can provide. But even as Bowman acknowledges that "the cycle could be turning somewhat," it's not at all clear what such a turn might mean. In their paper "The Evolution of Regional Seismicity between Large Earthquakes," he and Geoff King assume a highly skeptical position: "We emphasize that the use of this technique as a method of earthquake prediction should be approached with caution," they write, "since the method requires forehand knowledge of the fault or fault segments that will rupture in a future event." What this suggests is that, even in a regional setting, prediction remains, at best, a crapshoot, a matter of luck, of probabilities, as amorphous as it's always been. "The problem," Bowman says, "is that we're limited by our understanding of geology. We have to recognize where an earthquake might be before it happens. Even if we'd had a prediction system, for instance, we would have never gotten Landers, because nobody thought a single earthquake could link together three faults. We would have never known to look for that."

"And Hector Mine?" I ask, thinking about the Lavic Lake Fault, and how no one except Cloud Man saw that earthquake coming, or even thought an earthquake could be possible there at all.

"No," Bowman tells me. "We would have never gotten Hector Mine."

A couple of days after my visit to Bowman, I'm going through a stack of papers in my office when I find Cloud Man's photographs. One minute I'm sitting on the floor, cross-legged, looking at interview notes and USGS reports, and the next, I'm holding a small green-and-white Fuji Film folder with the name and address of Shou's daughter etched in calligrapher's pencil on the outside, and five color snapshots nestled within. Although I've always known I had these pictures somewhere, I couldn't be more surprised to see them than if we'd just had another earthquake on a ten-thousand-year-dormant fault.

I stand up and remove the photos from their packet, spread them in a loose arc across the surface of my desk. It's been a couple of years since I've last seen them, but they are exactly as I remember, all snapped from a distance, and timed and dated in

Cloud Man's neat hand on the back. A shot taken the morning of October 18, 1994, reveals a cloud like a feather in the blank blue sky east of Pasadena; it scuds over suburban rooftops and points down vaguely, suggesting an epicenter somewhere behind the palm trees that blossom along the bottom of the frame. In another picture, a similar shape rises from a thin layer of cirrus clouds, reminding me of nothing so much as Olga Kolbek's Old Faithful geyser in full eruption, a glorious spurt of precipitation exploding through the middle of the air. The most intriguing image is the cloud Shou claims predicted Northridge, which he photographed on January 8, 1994, nine days before the earthquake, although looking at it today, I'm mostly struck by its indistinction, the way it shows me so little, just a faint white line high in the sky that mostly resembles a smudge. Again, I wonder, how do you read this? How does it give you an earthquake like Northridge, or, for that matter, Hector Mine? I think back to the cloud I saw at Berkland's, to its piercing specificity, a specificity that's missing in these photographs. It's the best example of an earthquake cloud I've ever come across, and yet it did not yield an earthquake, at least not any earthquake that I know. A few months from now, I will see another such cloud while walking along Robertson Boulevard, a cloud with almost the exact same shape, like a quill pen, dangling out of the sky to the southeast. I will point it out to Rae, who will be walking with me; she will laugh and say, "I wonder where the earthquake's going to be." Later that day, we will hear about a 5.3 in the Imperial Valley, not too far from San Diego, exactly where that cloud appeared. And in that moment, I will once again ask myself those questions, uneasy with anxiety and awe.

This afternoon, though, all that's in the future, just as these pictures seem to be wholly of the past. For a moment, I realize that I can get in touch with Shou now if I want to, but just as quickly, I understand that I never will. Instead, I take his photos and place them in an envelope, address it, and slap on a couple of stamps. Then, I step outside, into a cloudless day, the air a symphony of avian arpeggios, and walk the five blocks to the post office to drop them in the mail.

THE MYTH OF SOLID GROUND

Let me tell you about the most recent earthquake I remember. It happened not long ago, on a Saturday in January, after Noah and I had been out riding the Metro Red Line, sitting in the front car, as we generally do. That day, we had no particular destination; we had been drawn, instead, by the idea of going somewhere, of riding to a different part of the city, of transforming the indistinction of a winter afternoon. This is hardly an uncommon occurrence, for despite my discomfort at being below street level in a seismic zone, Noah loves to ride the subway in Los Angeles, as he does most anywhere you might name. We have spent whole days in the Bay Area, for instance, traveling back and forth between San

Francisco and Oakland, and once, on a bitter December New York morning, we took the Number 7 train from Grand Central Station to Willets Point, just for the experience of traversing an urban territory on an elevated track. In Southern California, however, the subway is a far more limited operation, and if you don't want to go to Universal City, the only other option is downtown.

Noah and I drove to Hollywood, to the Hollywood and Vine station, where we parked and took the escalator down. At the bottom of that moving staircase, before we passed through the vaulted corridor towards the platform, I looked up at the last faint trace of sunlight flickering above us, like a beacon or a warning sign. This is the moment that always gets me, the moment of decision, the moment of no return. It's like the instant the flight attendant closes the door of the airplane; no matter what, you're in it now, at the mercy of forces greater than yourself. The same, of course, is true of every second you spend in earthquake country, but it's only in spots like this—a subway station or a parking structure or a high-rise building—that you confront such a reality head-on. There are no illusions underground, nothing to distract you, just these tunnels, carved out like catacombs, and your own half-articulated hope that, for the next few hours anyway, seismicity will remain a matter of geologic, as opposed to human, time.

I fed a string of dollar bills into the ticket machine, then Noah and I moved away from the sunlight, down into the embrace of the earth. We passed a display featuring a vintage movie camera, all black metal and 1930s rounded edges, a cross between an art deco sculpture and a functional machine. Like the Lankershim station, like every subway station in Los Angeles, this one, too, had been built to reflect a certain motif, an aspect of the city's personality; here, the theme was Hollywood, although not the real Hollywood, but a stylized ideal. The arched dome of the station ceiling was a mosaic of empty film reels, painted black to reflect the image of the sky at night. Down on the platform, support pillars were grooved to look like palm trees, each post exploding at the apex in a spray of fronds. The bench backs had been designed to resemble period cars—Packards, Buicks, Cadillacs, long and streamlined, with boxy chassis and extended, snoutlike hoods. On the boulevard above us, tourists gawked at the true Hollywood: junkies and hookers parading in front of Musso & Frank and the Pantages Theater like an inverse image of the glamour life. The first time I ever came here, I was stunned by this, by the way the reality so contradicted the illusion, the way the neighborhood—the streets, the people, the very atmosphere—was so completely antithetical to the myth. This, of course, is the story of Los Angeles, of California, the story of earthquakes and seismicity, as well. Everywhere I turn, there are overlays and underlays, narratives and threads and bits of inference, adding up to a chaotic landscape in which the deeper I look, the more complex all these interactions seem.

Noah and I took the first train downtown, rocketing through the narrow barrel of the tunnel as if we had been placed in a bullet being fired from a gun. Between stations, I stared out the front window, calculating distances, looking for the glimmer of approaching platforms, working and reworking my silent mantra: Please, not here, not now. Scattered throughout the car behind us, a dozen or so people sat reading, talking, studying maps of the system; looking at them, I considered again the leap of faith required to lay down roots in a seismic region, the way we must come to terms with our doubts, our fears, our most awful divinations, if we are to get on with the business of living at all. If you really thought about it, there was nothing but trouble to be gained down here, encircled by the earth as if by a living presence, feeling its heat, its weight, its thick and loamy stillness, like a promise that would one day be fulfilled. I turned to Noah, who had his face pressed up against the glass and was calling out the stations—Beverly/Vermont, Wilshire/Vermont, MacArthur Park—marking off the route to Pershing Square. Please, I thought again, not here, not now.

I don't remember very much about what Noah and I did once we emerged from the subway that January Saturday, just a few scattered impressions here and there. I know that we had lunch, and that we walked around Grand Central Market; I also know that during the course of the afternoon, Noah and I slid up the hill to California Plaza and watched the dancing waters on display. There was a stage there, a flat peninsula of pavement nestled between high-rise buildings, and I applauded as Noah did an improvised song-and-dance routine. Eventually, we made our way back down to Hill Street, where we found the subway entrance and descended to the platform, which, in this station, was overhung with dangling neon sculptures—stripes of red and blue and white light arranged in geometric patterns—as if this were less a transit system than a subterranean museum.

It was on the way back that I felt it. Not the earthquake, but something more disquieting—a premonition, a hallucination, a harbinger of a particularly vivid kind. I'm not sure where we were, under Vermont Avenue most likely, but I recall a rush of . . . not relief, but the *anticipation* of relief, the recognition that this trip was nearly over, that we would soon be back above the ground. I've often had such a reaction at the end of a ride on the L.A. subway, a sense of letting go, of existing slightly out of time. It's as if, so close to the finish, the minutes grow elastic, undergoing a critical phase transition in which each instant expands to contain its own past and future, and time becomes less a progression than an unending state of being. That day, however, things were different, for no sooner had I relaxed than the entire fabric of reality shifted, breaking through the surface of a placid Saturday to reveal the explosive possibility that lay beneath. Suddenly, my vision narrowed, and whatever had been on the periphery dropped away. All I could see now was the subway tunnel, unfurling in the train's front window like a three-dimensional feedback loop. As I watched, what appeared to be shock waves started to ripple towards us like enormous cartoon vibrations, echoing one after the other along the tunnel's length. The concrete embouchement seemed to shudder, and I could almost see the walls begin to crumble, see cracks propagate across those smooth and seamless surfaces like so many microcosmic faults. So this is what will happen, I found myself thinking, so this is how it will be. Then, the shock waves reached us, and, just like that, they dissipated, leaving me to gasp for breath. As the subway passed through to the other side, I looked around the car, but no one else had noticed—not even Noah, who stood before the window with me, hands wrapped around an imaginary steering wheel, pretending to drive the train.

Noah and I left the subway at Hollywood Boulevard, and rode the escalator up to find our car. While it felt good to be back in the daylight, I couldn't shake the sense of what I'd seen. In all the years I'd been involved with earthquakes, I had never had an experience like that. I had wanted to, had wanted to connect with seismicity at the deepest level; I had waited for tremors, anticipated tremors, dreamt of them. Now, though, I wasn't sure what to make of this, what, if anything, it might mean. The whole way back to our house, I could feel it as a low-grade pulsing, a buzzing in the back of my brain. What it brought back was the experience of Rae's last prediction, the almost unbearable edge of expectation, of time cast in suspension, of having been offered a glimpse into the secret workings of the world. Was this,

I asked myself, one of those moments? Had I been given something—a gift of second sight, of augury? Or was it just a fantasy, the work of an overactive imagination? I couldn't decide what I believed.

At home, I played with Noah and Sophie while Rae made dinner, or maybe we watched TV. Again, the details elude me, refuse to coalesce, as if we're talking about both myth and memory. Here, though, is what does not elude me, what I recall as distinctly as if it were happening this very instant, as distinctly as those coffee cups dancing on their shelves in San Francisco, as distinctly as Northridge or Hector Mine. It was after six when we sat down, the sky outside a blue-black sheen of darkness, empty of either clouds or stars. As we ate, Rae and the kids chattered back and forth across the table, while I kept quiet, my mind a chaos of uncertainty. Then, somewhere before the end of the meal, I felt a slight pitch and yaw, like a hiccup in the floor beneath me, and the whole house started to rock, gently, even easily, as if the world had been cast on rollers and was being shaken by a giant hand. "What?" I said, half-aloud, but it was just a reflex; I knew precisely what was going on. The tremor wasn't big—it was over in a few seconds, just a seismic whisper, barely long enough for us to look up before it disappeared under the night. Yet even in the midst of such a minor shaking, I could sense it echoing inside me, as I thought about those shock waves, and wondered what, exactly, I had witnessed underground.

The earthquake story I've just told you didn't really happen. Or, at least, it didn't really happen like that. Yes, I had some kind of reverie or vision on the Red Line, followed, at 6:26 the same evening, by a 4.3 temblor that, for all intents and purposes, shared an epicenter with the 1971 San Fernando quake. Yes, it was a Saturday in January, and yes, Noah was with me, both on the subway and during the tremor itself. What isn't true is that this is the most recent earthquake I remember, for the event I've just described took place in 2001, which means that as many as forty thousand earthquakes—the vast majority of them, obviously, too small to notice—have rolled through Southern California since then. In all the years the USGS has monitored seismicity in the state, only once has a twenty-four-hour period passed without any recorded ground motion, and this was such an anomaly that, at first, Survey seismologists thought their equipment had to be at fault. Even on a conscious level, I can bring to mind at least two or three subsequent earthquakes, all minor, 2.8s and 3.1s, with nothing in particular to distinguish them, to make them stand apart. Yet lest this seem gratuitous, as if I'm telling a lie simply for the sake of narrative, let me assure you that there's more at stake. The earthquake I've just recounted may not be the last one I remember, but it is the one that (in the years since Hector Mine, anyway) I remember *most*. It's the one that lingers, the one that resonates; it's the one that seems to hint at something, some kind of inexplicable wildness, the way that, here in California, we are always coming up against the intersection of different layers, different aspects of reality. In that sense, I raise it not just because it's a strange story, but because in its strangeness, it may tell us something about earthquakes, about how they get inside us—in other words, how we live with them.

Of all the issues such an earthquake raises, the most fundamental may have to do with prediction—or, more accurately, with prediction and its discontents. By saying this, I don't mean to suggest that my experience was prophetic; to tell the truth, I don't know what happened on that subway, and I don't think it matters anyway. Whether it was a premonition or a coincidence, a fluke or some strange extrasensory occurrence, what's important is the emotional tenor of the incident, its inability to console me, to set my heart or mind at ease. I spent the hour or two after those shock waves in a state of slow and subtle agony, questioning what I'd seen and all its possible meanings, equally desiring and not desiring the fulfillment of the quake. This is the dirty secret of prediction, the way it makes us long for earthquakes, if only to bestow a shape, a perspective, on what we think we know. Ultimately, though, that's just one more chimera, a prayer out of the darkness for control. Earthquakes, after all, are too intractable, too chaotic, which renders prediction too . . . *unpredictable,* a source of anxiety even (or especially) when it works. How, for instance, do you reckon with a premonition? And what if it doesn't come to pass? No, in the end, prediction is too narrow a window, a strategy for framing seismicity through a human filter, which is why it continues to elude us, having more to do with what we want from earthquakes than what earthquakes can provide us on their own.

The idea of framing seismicity through a human filter emerges again and again in our interactions with earthquakes, from the deepest level to the most mundane. If you go up on the Web site of the Metro Red Line, you'll find a few paragraphs on "Earthquake and Fire Safety" that address this conundrum in entirely reassuring terms. "Though many question the existence of a subway in 'Earthquake Country,' " the site tells us, "few actually realize that there are many other cities in the world with subways and seismic activity. San Francisco, Tokyo and Mexico City, for example. All of their respective systems suffered very little or no damage as a result of a devastating earthquake." It's tempting to be cynical about such a statement, especially when it's preceded by a note informing us that the subway is "one of the safest places to be" in an earthquake—because "there are no structures or falling objects to endanger you." But while I'd be the last to ascribe pure motives to the Los Angeles Metropolitan Transit Authority, I can't help thinking that all of this—not the statements per se, but the sensibility that informs them—is less a matter of *de*lusion than *il*lusion, a way of making peace with our existence in a seismic zone. How, after all, do we live with earthquakes? That is still the question, after all this time. For many of us, the key is not to think about it, or to think about it in ways that minimize the enormity, that reduce the subject to a level we can comprehend. Prediction is one end of that spectrum; at the other, we find jokes and games, curios and knickknacks, all the little tricks we use to hold seismicity at bay. In my office, I have a tin can labeled "California Canned Earthquakes" that has a battery inside, and when you turn it over, it begins to vibrate, simulating the movement of a 6.5. My kids play with a book called *Earthquake!* that features, in a small box on the cover, a three-dimensional cityscape with a pull string—yank the cord, and the city shakes. There are at least two California sports teams called the Earthquakes, just as New England has clubs called the Revolution and the Patriots. A downtown Los Angeles restaurant named Epicentre even serves a San Andreas soup, in which light and dark beans are separated by a line that mirrors the angle of the fault.

Perhaps the most extreme example of this attitude I've ever come across is in the town of Frazier Park, a small mountain hamlet just off the I-5 near the crest of Tejon Pass. More than most places in California, Frazier Park occupies the center of the seismic bull's-eye, and nowhere is that more apparent than at Frazier Mountain High School, which sits on a hill overlooking the freeway, literally atop the intersection of the San Andreas and the Garlock faults. Because of its location, the school has been constructed like a fortress, squat and square and slablike, an enormous concrete bunker with barely a dusting of windows set within its thick gray walls. Still, when, one afternoon last fall, I paid a visit, no one seemed to know about the faults at all. "There are earthquake faults here?" asked a woman in the administration office, while the school custodian told me that, even though the building had been designed according to rigorous seismic guidelines, the threat was overstated because the faults were not so very close.

"Really?" I said. "I'd heard they were right here."

"No." He waved me off with a loose sweeping gesture. "They're down the hill a ways. Not here."

"How far down the hill?"

The custodian was silent for a moment. I could see his eyes recede, as if he were computing something—distances or odds. "I don't know," he said, finally. "Maybe five hundred feet."

There are a couple of ways to consider such a comment. On the one hand, it's a bad joke, an example of how little we understand about the world around us, about its dynamics, its dangers, both real and implied. At the same time, that custodian is speaking for all of us who live in California, where, metaphorically or actually, *everything* is five hundred feet from the fault. In that sense, what he's offering is a psychological defense mechanism, a survival tactic, a seismological bottom line. Yes, he's saying, we all know this is an elemental landscape, but in the face of that, you do what you can, and then you walk away and try not to worry about it anymore. Still, as much as I understand the impulse—who wants to spend all their time obsessing over earthquakes?—it's a strategy that almost certainly will come back to haunt us, a bit of willful ignorance, or even hubris, our own collective tragic flaw. Frazier Park, after all, sits in the heart of the San Andreas segment that broke in 1857, a segment that David Bowman, for one, has called the most likely place for another major earthquake on the fault. Other researchers point to the

San Bernardino section of the San Andreas, the very section I once climbed down into, at least part of which hasn't ruptured in almost 325 years. "We already knew we had a high seismic risk in the Inland Empire," Sally McGill said recently in the *Los Angeles Times*. "This isn't a call for new alarm." And yet, if not alarm, it does feel like a call for something, for awareness, for engagement, for a different way of thinking about seismicity and the depth at which it runs throughout our lives.

The irony is that this new way of thinking is right before us, inscribed like some ancient language in the ground. Earthquakes are the icons of that language; they are its alphabet, its punctuation, its sentences and paragraphs, the mythology it adds up to, the very tales it tells. As for why this is . . . well, earthquakes work, if we look at them correctly, if we attempt to understand them in their own time. They influence one another, they interact with one another, they connect in a complex system composed in equal parts of order and chaos, of what we can expect and what we can never expect, of all the things we know and do not know. Perhaps the easiest way to put it is to say that while we know there are relationships, they are relationships on which we can't quite get a handle, except in the most general terms. We can say that a quake is coming, but not exactly where or when. We can say that strain is building, that there seems to be precursory activity in a region, that seismicity is mounting or receding, that we're in a seismic lull. Still, even in the most promising precincts, earthquakes have a way of confounding us, of appearing when they're least expected, or not appearing at all. In the end, it all boils down to something David Bowman told me when I visited him in Fullerton. "Science," he said, "is not about answers. It's about questions. After you know an answer, it's not science anymore."

It is this middle ground we inhabit in the realm of earthquakes, a place where the questions *become* the answers—or, more important, where these distinctions do not matter any more. Once, maybe a year ago, I tried to explain this to a friend, although at the time, I could not find the words. We had been discussing the structure of seismicity, the way it simultaneously reveals and conceals itself, and he was pressing me about what, exactly, I believed. "I believe in an orderly universe," I finally said. But that's not it, not really, not at the most essential level, anyway. No, the truth is that I also believe in chaos, in wildness, in the idea that there's a meaning here I can't quite see. I believe in order, in other words, but in an order that eludes me, that exists a little bit beyond my reach. This is what compels me about California, the idea that here, even the most basic assumptions contain their own uncertainty. But it is also one of the most powerful solaces that earthquakes offer, a profound and lasting sense of mystery.

Of course, in order to enter such a landscape, we need to walk away from many things. We need to walk away from control, from the idea that order is something we can see. We need to walk away from ourselves, from our narrow view of time, of geology. To live with earthquakes is to have one foot in the present and the other in the deepest reaches of the past. It is to find a balance, to understand that everything is always up for grabs. If geophysicists like Ross Stein and David Bowman are right—even if they're not—the earth, as a living planet, is always changing, which means there's no such thing as solid ground. It's another myth, as mutable as magma, a Pangaea of the mind. In a hundred million years, none of what we see will be here, and I'm not talking only about human life. I mean the mountains, the oceans, the very continents, the shape and texture of the world. In a very real way, this cannot help but diminish us, by reducing our lives to a biological afterthought, an ephemeral sideshow to the geologic drama, the fluid dance of plate tectonics, the interplay of movement along faults. Yet if we open our minds a little and look at it a different way, we may, paradoxically, find ourselves enlarged. Human life, after all, is only ephemeral on its own terms, when we consider it independent of everything else. Place it in a context, in conjunction with something larger, and it immediately takes on a more complicated shape. This is the secret of religion, which seeks to frame us beneath the mantle of some amorphous eternity. When it comes to earthquakes, though, eternity is right in front of us, if we allow ourselves to see.

How do we manage this? The truth is, I don't know. But let me suggest one small possibility, a way of living in the presence of earthquakes, albeit on the most personal of scales. The night of that January temblor, after the shaking had subsided, Noah and Rae and Sophie and I all remained silent at the table, waiting for our internal flutters to recede. Finally, in a small voice, somewhat hesitant, Noah asked, "What was that?"

"That was an earthquake," I said, and waited to see how he'd respond.

For a moment, Noah didn't do anything, just mulled it over, his face a shifting storm of moods. I could see a lot of thoughts there, a lot of emotions—confusion, fear, even a whisper of prognostication—as he tried to figure out what he felt. Eventually, though, his eyes began to brighten, and his mouth unfolded in a grin. "An earthquake?" he repeated.

I nodded.

"Cool," Noah said.

INDEX